Lecture Notes in Mathematics   2220

More information about this series at http://www.springer.com/series/304

Fondazione C.I.M.E., Firenze

FONDAZIONE CIME
ROBERTO CONTI
CENTRO INTERNAZIONALE MATEMATICO ESTIVO
INTERNATIONAL MATHEMATICAL SUMMER CENTER

C.I.M.E. stands for *Centro Internazionale Matematico Estivo*, that is, International Mathematical Summer Centre. Conceived in the early fifties, it was born in 1954 in Florence, Italy, and welcomed by the world mathematical community: it continues successfully, year for year, to this day.

Many mathematicians from all over the world have been involved in a way or another in C.I.M.E.'s activities over the years. The main purpose and mode of functioning of the Centre may be summarised as follows: every year, during the summer, sessions on different themes from pure and applied mathematics are offered by application to mathematicians from all countries. A Session is generally based on three or four main courses given by specialists of international renown, plus a certain number of seminars, and is held in an attractive rural location in Italy.

The aim of a C.I.M.E. session is to bring to the attention of younger researchers the origins, development, and perspectives of some very active branch of mathematical research. The topics of the courses are generally of international resonance. The full immersion atmosphere of the courses and the daily exchange among participants are thus an initiation to international collaboration in mathematical research.

C.I.M.E. Director (2002 – 2014)
Pietro Zecca
Dipartimento di Energetica "S. Stecco"
Università di Firenze
Via S. Marta, 3
50139 Florence
Italy
e-mail: zecca@unifi.it

C.I.M.E. Director (2015 –)
Elvira Mascolo
Dipartimento di Matematica "U. Dini"
Università di Firenze
viale G.B. Morgagni 67/A
50134 Florence
Italy
e-mail: mascolo@math.unifi.it

C.I.M.E. Secretary
Paolo Salani
Dipartimento di Matematica "U. Dini"
Università di Firenze
viale G.B. Morgagni 67/A
50134 Florence
Italy
e-mail: salani@math.unifi.it

CIME activity is carried out with the collaboration and financial support of INdAM (Istituto Nazionale di Alta Matematica)
For more information see CIME's homepage:
http://www.cime.unifi.it

Xavier Cabré • Antoine Henrot •
Daniel Peralta-Salas • Wolfgang Reichel •
Henrik Shahgholian

Geometry of PDEs and Related Problems

Cetraro, Italy 2017

Chiara Bianchini, Antoine Henrot, Rolando Magnanini
Editors

 Springer

Authors

Xavier Cabré
Departament de Matemàtiques
ICREA and Universitat Politècnica de Catalunya
Barcelona, Spain

Daniel Peralta-Salas
Consejo Superior de Investigaciones Científicas
Instituto de Ciencias Matemáticas
Madrid, Spain

Henrik Shahgholian
Department of Mathematics
The Royal Institute of Technology
Stockholm, Sweden

Antoine Henrot
Institut Elie Cartan – Faculté des Sciences et
Technologies
Université de Lorraine
Vandoeuvre-les-Nancy, France

Wolfgang Reichel
Institute for Analysis, Department of Mathematics
Karlsruhe Institute of Technology
Karlsruhe, Germany

Editors

Chiara Bianchini
Dipt di Matematica e Informatica Ulisse Dini
Università degli Studi di Firenze
Firenze, Italy

Antoine Henrot
Institut Elie Cartan – Faculté des Sciences et
Technologies
Université de Lorraine
Vandoeuvre-les-Nancy, France

Rolando Magnanini
Dipt di Matematica e Informatica Ulisse Dini
Università degli Studi di Firenze
Firenze, Italy

ISSN 0075-8434   ISSN 1617-9692   (electronic)
Lecture Notes in Mathematics
ISBN 978-3-319-95185-0   ISBN 978-3-319-95186-7   (eBook)
https://doi.org/10.1007/978-3-319-95186-7

Library of Congress Control Number: 2018953697

Mathematics Subject Classification (2010): Primary 35Jxx; Secondary 35B06, 35B10, 35J20, 35P15, 35P30, 35G15, 35N25, 35R01, 35R11, 35R35, 35R37, 35Q79, 35Q91, 35Q92, 58J32.

This Springer imprint is published by the registered company Springer Nature Switzerland AG
The registered company address is: Gewerbestrasse 11, 6330 Cham, Switzerland

Preface

This volume collects together the lecture notes of the five minicourses given at the CIME Summer School "Geometry of PDEs and related problems," held in Cetraro (Cosenza, Italy) in the week of June 19–23, 2017.

The lectures were given by Xavier Cabré (ICREA and Universitat Politècnica de Catalunya, Barcelona), Antoine Henrot (Université de Lorraine, Nancy), Daniel Peralta-Salas (ICMAT, Instituto de Ciencias Matemáticas, Madrid), Wolfgang Reichel (Karlsruher Institut für Technologie, Karlsruhe), and Henrik Shahgholian (Royal Institute of Technology, Stockholm).

The aim of the school was to present different aspects of the deep interplay between partial differential equations and geometry. Certainly, PDEs are classically a powerful tool for the investigation of important problems coming from geometry (and the applied sciences). On the other hand, it is also unquestionable that geometry gives useful and often decisive insights into the study of PDEs and related problems. Now that basic questions about PDEs—such as existence, uniqueness, stability, and regularity of solutions for initial/boundary value problems—have been fairly well understood, research on topological and/or geometric properties of their solutions have become more vigorous. The summer school aimed to present an overview of a few themes of recent research in the field and their mutual links, describe the main underlying ideas, and provide up-to-date references.

The five lecture notes in this book present in a friendly manner a broad spectrum of results, describing the state-of-the-art in this area of research. They also give further details on the main ideas of the proofs, their technical difficulties, and their possible extension to other contexts. Last but not least, a collection of open problems can be found scattered here and there in the volume.

In what follows, we shall attempt to give a brief description of the contents of the five essays that could work as an invitation to the reading of this volume.

We start with the paper "Stable Solutions to Some Elliptic Problems: Minimal Cones, the Allen–Cahn Equation, and Blow-up Solutions," written by X. Cabré with the collaboration of G. Poggesi. Its focus is on the classification of stable solutions to some nonlinear elliptic equations. This topic still attracts considerable attention, after more than half a century, in the wake of the seminal works of J. Simons,

E. Bombieri, E. De Giorgi, and E. Giusti in the late 1960s. The main ideas are illustrated by considering three different equations, emphasizing that the techniques involved in the three settings are quite similar. As a starter, a classical result is presented. In fact, the minimality of the Simons' cone in high dimensions is proved, and Simons' proof of the flatness of stable minimal cones in low dimensions is given in almost full detail. A semilinear analogue pertains to a conjecture posed by E. De Giorgi in 1978 for solutions of the Allen–Cahn equation in Euclidean space. For this second problem, an insightful account of selected relevant results is presented. Also discussed is an open problem in high dimensions on the saddle-shaped solution vanishing on the Simons' cone. The third problem concerns a question posed by H. Brezis around 1996 on the boundedness of stable solutions to reaction–diffusion equations in bounded domains. Proofs of their boundedness in dimension up to 4 are presented, and the main open problem of their boundedness in dimensions 5 to 9 is discussed. Brief comments on related results for harmonic maps, free boundary problems, and nonlocal minimal surfaces are also given throughout the paper.

The lecture notes "Isoperimetric Inequalities for Eigenvalues of the Laplacian," by A. Henrot, give an overview of sharp inequalities for the Dirichlet eigenvalues of the Laplace operator subject to geometric constraints on the underlying domain. Such constraints can be on the volume or perimeter of the domain or may involve box constraints or, yet, restrictions to some specific subclasses like polygons or convex sets.

The different constraints are first examined in the case of the first Dirichlet eigenvalue. To begin with, the classical Rayleigh–Faber–Krahn inequality is presented together with a (recent) quantitative version of it. The same minimization problem in the subclass of polygons is discussed, and the famous Pólya conjecture, which states that the regular polygon has the least first eigenvalue among all the polygons with the same number of sides and given area, is presented. Finally, some results on such minimization problems with a box constraint are reported, and the problem of how to place an obstacle in a domain in order to optimize its first eigenvalue is discussed. Concerning the second eigenvalue, a proof of the Hong–Krahn–Szegö theorem is presented (with a quantitative version). This result states that the least second eigenvalue for bounded domains with given volume is achieved by the union of two identical balls. The problem of minimizing the second eigenvalue among convex domains is then considered. The essay closes by reporting on minimization problems for higher-order eigenvalues subject to volume, perimeter, or diameter constraints.

Throughout the discussion, classical tools of modern analysis and shape optimization are recalled. They pertain to rearrangement techniques such as Schwarz and Steiner symmetrizations, Hausdorff convergence of compact and open sets, γ-convergence and convergence of eigenvalues, and domain derivative. The essay also proposes nine challenging open problems.

The paper "Topological Aspects of Critical Points and Level Sets in Elliptic PDEs," by D. Peralta-Salas and A. Enciso, has a stronger topological flavor. The authors discuss the emergence and complexity of topological structures in the study of the critical points and level sets of solutions to elliptic PDEs. This theme also has an important impact on the study of dynamical systems, since the curves of steepest descent of the relevant solutions can be thought of as trajectories of a related dynamical system.

A first result is of a global nature: the number of critical points of a Green's function on a surface of finite type admits a topological upper bound. This property is shown not to hold in higher dimensions. A second, somewhat surprising, result shows by construction that solutions of elliptic PDEs may have level sets with very complicated topologies. The construction is based on Thom's isotopy theorem and a powerful Runge-type global approximation theorem created by the authors. The method is illustrated by considering solutions of the Helmholtz equation. Applications of these ideas are the construction of Schrödinger operators in Euclidean space with eigenfunctions having nodal lines of arbitrary knot type and of bounded solutions to the Allen–Cahn equation with level sets of any compact topology. The former result gives a positive answer to a 2001 conjecture of Sir Michael Berry.

In his essay, "Symmetry Properties for Solutions of Higher-Order Elliptic Boundary Value Problems," W. Reichel touches upon the question of whether the solutions of nonlinear boundary value problems inherit the same symmetries as the underlying domain. In the case of second-order elliptic equations, this has been an area of intense research in the last half century, starting from the pioneering works of J. Serrin; H.F. Weinberger, B. Gidas, W.M. Ni, and L. Nirenberg. For higher-order cases, few results are available.

The essay presents three examples—that involve linear and semilinear polyharmonic equations either on the Euclidean space or on balls with additional Dirichlet boundary conditions—in which the desired symmetry can be obtained by using different methods from nonlinear analysis and geometry. The three methods are, respectively, based on a fixed point argument via the contraction mapping theorem, the method of moving planes (in the wake of Serrin's approach), and a method (reminiscent of that of Weinberger) based on integral identities and inequalities and Newton's inequalities for the symmetric functions of the eigenvalues of a symmetric matrix.

The volume ends with the article "Recent Trends in Free Boundary Regularity," by H. Shahgholian. Here, the center of attention is the presentation of a number of problems within the applied sciences having free boundaries as a common denominator. A free boundary problem emerges when one or several functions, typically solutions of differential equations or systems, must be determined in a priori unknown domains of the space, or space-time. In a large class of these problems, the process is usually controlled by certain constraints governing a phase transition. Phase transitions (and the associated free boundaries) are detected in a wide spectrum of applications, spanning physical, economical, financial, and biological phenomena where a qualitative change of the medium takes place. Evident phase transitions occur when ice changes to water or a liquid to crystal. Less evidently, phase transitions can be envisioned in processes involving buying and selling assets or active and inactive biological populations, to name a few.

The essay starts with a brief look at how five important free boundary problems can be modeled. Next, there is a closer look at the main mathematical questions (and the related techniques) of interest in the study of free boundaries. This is done by using as a bench test the mathematical theory of obstacle problems. For these problems, questions like existence, optimal regularity, non-degeneracy, Hausdorff dimension, monotonicity formulas, global solutions, Lipschitz, and C^1-regularity of free boundaries are discussed with a considerable level of detail. The rest of the paper is dedicated to a variety of other types of free boundary problems (classical, like Bernoulli's problem, and not like nonlocal problems) and recent trends and developments.

We hope that these lecture notes will find the appreciation of scholars and students and contribute to the enhancement of research in these stimulating fields.

We want to express our gratitude to the five speakers of the CIME Summer Course for their excellent work during the course and for their inspiring essays written for this volume. We also wish to thank PhD students Diego Berti, Nico Lombardi, Giorgio Poggesi of the Università di Firenze, and Lorenzo Cavallina of Tohoku University (Sendai, Japan) for their effective help in editing this book.

We finally acknowledge, in addition to the CIME Foundation's support, the financial contribution of the MIUR-FIR2013 project *Geometrical and Qualitative aspects of PDEs* and of the Progetto Strategico di Ateneo *Equazioni differenziali: aspetti geometrici, disuguaglianze collegate e applicazioni* of the Università di Firenze.

Firenze, Italy   Chiara Bianchini
Nancy, France   Antoine Henrot
Firenze, Italy   Rolando Magnanini

Contents

Chapter 1
Stable Solutions to Some Elliptic Problems: Minimal Cones, the Allen-Cahn Equation, and Blow-Up Solutions

Xavier Cabré and Giorgio Poggesi

Abstract These notes record the lectures for the CIME Summer Course taught by the first author in Cetraro during the week of June 19–23, 2017. The notes contain the proofs of several results on the classification of stable solutions to some nonlinear elliptic equations. The results are crucial steps within the regularity theory of minimizers to such problems. We focus our attention on three different equations, emphasizing that the techniques and ideas in the three settings are quite similar.

The first topic is the stability of minimal cones. We prove the minimality of the Simons cone in high dimensions, and we give almost all details in the proof of J. Simons on the flatness of stable minimal cones in low dimensions.

Its semilinear analogue is a conjecture on the Allen-Cahn equation posed by E. De Giorgi in 1978. This is our second problem, for which we discuss some results, as well as an open problem in high dimensions on the saddle-shaped solution vanishing on the Simons cone.

The third problem was raised by H. Brezis around 1996 and concerns the boundedness of stable solutions to reaction-diffusion equations in bounded domains. We present proofs on their regularity in low dimensions and discuss the main open problem in this topic.

Moreover, we briefly comment on related results for harmonic maps, free boundary problems, and nonlocal minimal surfaces.

The abstract above gives an account of the topics treated in these lecture notes.

X. Cabré (✉)
Universitat Politècnica de Catalunya, Departament de Matemàtiques, Barcelona, Spain

ICREA, Pg. Lluis Companys 23, Barcelona, Spain
e-mail: xavier.cabre@upc.edu

G. Poggesi
Università di Firenze, Dipartimento di Matematica ed Informatica "U. Dini", Firenze, Italy
e-mail: giorgio.poggesi@unifi.it

© Springer Nature Switzerland AG 2018
C. Bianchini et al. (eds.), *Geometry of PDEs and Related Problems*,
Lecture Notes in Mathematics 2220, https://doi.org/10.1007/978-3-319-95186-7_1

1.1 Minimal Cones

In this section we discuss two classical results on the theory of minimal surfaces: Simons flatness result on stable minimal cones in low dimensions and the Bombieri-De Giorgi-Giusti counterexample in high dimensions. The main purpose of these lecture notes is to present the main ideas and computations leading to these deep results—and to related ones in subsequent sections. Therefore, to save time for this purpose, we do not consider the most general classes of sets or functions (defined through weak notions), but instead we assume them to be regular enough.

Throughout the notes, for certain results we will refer to three other expositions: the books of Giusti [28] and of Colding and Minicozzi [16], and the CIME lecture notes of Cozzi and Figalli [17]. The notes [13] by the first author and Capella have a similar spirit to the current ones and may complement them.

Definition 1.1 (Perimeter) Let $E \subset \mathbb{R}^n$ be an open set, regular enough. For a given open ball B_R we define the *perimeter of E in B_R* as

$$P(E, B_R) := H_{n-1}(\partial E \cap B_R),$$

where H_{n-1} denotes the $(n-1)$-dimensional Hausdorff measure (see Fig. 1.1).

The interested reader can learn from [17, 28] a more general notion of perimeter (defined by duality or in a weak sense) and the concept of set of finite perimeter.

Definition 1.2 (Minimal Set) We say that an open set (regular enough) $E \subset \mathbb{R}^n$ is a *minimal set* (or a *set of minimal perimeter*) if and only if, for every given open ball B_R, it holds that

$$P(E, B_R) \leq P(F, B_R)$$

for every open set $F \subset \mathbb{R}^n$ (regular enough) such that $E \setminus B_R = F \setminus B_R$.

Fig. 1.1 The perimeter of E in B_R

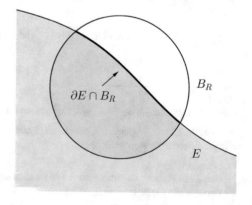

$\partial E \cap B_R$

B_R

E

Fig. 1.2 A normal
deformation E_t of E

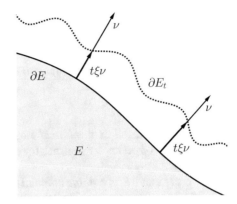

In other words, E has least perimeter in B_R among all (regular) sets which agree
with E outside B_R.

To proceed, one considers small perturbations of a given set E and computes the
first and second variations of the perimeter functional. To this end, let $\{\phi_t\}$ be a
one-parameter family of maps $\phi_t : \mathbb{R}^n \to \mathbb{R}^n$ such that ϕ_0 is the identity I and all
the maps $\phi_t - I$ have compact support (uniformly) contained in B_R.

Consider the sets $E_t = \phi_t(E)$. We are interested in the perimeter functional
$P(E_t, B_R)$. One proceeds by choosing $\phi_t = I + t\xi\nu$, which shifts the original set
E in the normal direction ν to its boundary. Here ν is the outer normal to E and is
extended in a neighborhood of ∂E to agree with the gradient of the signed distance
function to ∂E, as in [28] or in our Sect. 1.1.3 below. On the other hand, ξ is a scalar
function with compact support in B_R (see Fig. 1.2).

It can be proved (see chapter 10 of [28]) that the first and second variations of
perimeter are given by

$$\frac{d}{dt} P(E_t, B_R)\Big|_{t=0} = \int_{\partial E} \mathscr{H} \xi \, dH_{n-1}, \tag{1.1}$$

$$\frac{d^2}{dt^2} P(E_t, B_R)\Big|_{t=0} = \int_{\partial E} \left\{ |\delta\xi|^2 - (c^2 - \mathscr{H}^2)\xi^2 \right\} dH_{n-1}, \tag{1.2}$$

where $\mathscr{H} = \mathscr{H}(x)$ is the *mean curvature* of ∂E at x and $c^2 = c^2(x)$ is the sum of
the squares of the $n-1$ principal curvatures $k_1, \ldots, k_{n-1}$ of ∂E at x. More precisely,

$$\mathscr{H}(x) = k_1 + \cdots + k_{n-1} \quad \text{and} \quad c^2 = k_1^2 + \cdots + k_{n-1}^2.$$

In (1.2), δ (sometimes denoted by ∇_T) is the tangential gradient to the surface ∂E,
given by

$$\delta\xi = \nabla_T \xi = \nabla\xi - (\nabla\xi \cdot \nu)\nu \tag{1.3}$$

for any function ξ defined in a neighborhood of ∂E. Here ∇ is the usual Euclidean gradient and ν is always the normal vector to ∂E. Being δ the tangential gradient, one can check that $\delta\xi_{|\partial E}$ depends only on $\xi_{|\partial E}$. It can be therefore computed for functions $\xi : \partial E \rightarrow \mathbb{R}$ defined only on ∂E (and not necessarily in a neighborhood of ∂E).

Definition 1.3

(i) We say that ∂E is a *minimal surface* (or a *stationary surface*) if the first variation of perimeter vanishes for all balls B_R. Equivalently, by (1.1), $\mathcal{H} = 0$ on ∂E.

(ii) We say that ∂E is a *stable minimal surface* if $\mathcal{H} = 0$ and the second variation of perimeter is nonnegative for all balls B_R.

(iii) We say that ∂E is a *minimizing minimal surface* if E is a minimal set as in Definition 1.2.

We warm the reader that in some books or articles "minimal surface" may mean "minimizing minimal surface".

Remark 1.4

(i) If ∂E is a minimal surface (i.e., $\mathcal{H} = 0$), the second variation of perimeter (1.2) becomes

$$\left.\frac{d^2}{dt^2} P(E_t, B_R)\right|_{t=0} = \int_{\partial E} \left\{|\delta\xi|^2 - c^2\xi^2\right\} dH_{n-1}. \tag{1.4}$$

(ii) If ∂E is a minimizing minimal surface, then ∂E is a stable minimal surface. In fact, in this case the function $P(E_t, B_R)$ has a global minimum at $t = 0$.

1.1.1   The Simons Cone. Minimality

Definition 1.5 (The Simons Cone)   The Simons cone $\mathscr{C}_S \subset \mathbb{R}^{2m}$ is the set

$$\mathscr{C}_S = \{x \in \mathbb{R}^{2m} : x_1^2 + \ldots + x_m^2 = x_{m+1}^2 + \ldots + x_{2m}^2\}. \tag{1.5}$$

In what follows we will also use the following notation:

$$\mathscr{C}_S = \{x = (x', x'') \in \mathbb{R}^m \times \mathbb{R}^m : |x'|^2 = |x''|^2\}.$$

Let us consider the open set

$$E_S = \left\{x \in \mathbb{R}^{2m} : u(x) := |x'|^2 - |x''|^2 < 0\right\},$$

and notice that $\partial E_S = \mathscr{C}_S$ (see Fig. 1.3).

Fig. 1.3 The set E_S and the Simons cone $\mathscr{C}_S$

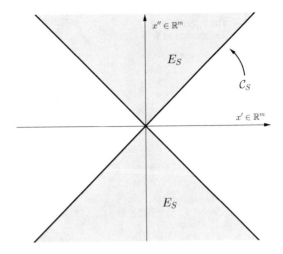

Exercise 1.6 Prove that the Simons cone has zero mean curvature for every integer $m \geq 1$. For this, use the following fact (that you may also try to prove): if

$$E = \{x \in \mathbb{R}^n : u(x) < 0\}$$

for some function $u : \mathbb{R}^n \to \mathbb{R}$, then the mean curvature of ∂E is given by

$$\mathscr{H} = \operatorname{div}\left(\frac{\nabla u}{|\nabla u|}\right)\bigg|_{\partial E}. \tag{1.6}$$

Remark 1.7 It is easy to check that, in $\mathbb{R}^2$, $\mathscr{C}_S$ is not a minimizing minimal surface. In fact, referring to Fig. 1.4, the shortest way to go from P_1 to P_2 is through the straight line. Thus, if we consider as a competitor in B_R the interior of the set

$$F := \overline{E}_S \cup \overline{T}_1 \cup \overline{T}_2,$$

where T_1 is the triangle with vertices O, P_1, P_2, and T_2 is the symmetric of T_1 with respect to O, we have that F has less perimeter in B_R than E_S.

In 1969 Bombieri, De Giorgi, and Giusti proved the following result.

Theorem 1.8 (Bombieri-De Giorgi-Giusti [5]) *If $2m \geq 8$, then E_S is a minimal set in $\mathbb{R}^{2m}$. That is, if $2m \geq 8$, the Simons cone $\mathscr{C}_S$ is a minimizing minimal surface.*

The following is a clever proof of Theorem 1.8 found in 2009 by G. De Philippis and E. Paolini ([23]). It is based on a *calibration argument*. Let us first define

$$\tilde{u} = |x'|^4 - |x''|^4; \tag{1.7}$$

Fig. 1.4 The Simons cone
$\mathcal{C}_S$ is not a minimizer in $\mathbb{R}^2$

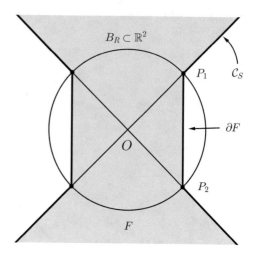

clearly we have that

$$E_S = \left\{ x \in \mathbb{R}^{2m} : \tilde{u}(x) < 0 \right\} \quad \text{and} \quad \partial E_S = \mathcal{C}_S.$$

Let us also consider the vector field

$$X = \frac{\nabla \tilde{u}}{|\nabla \tilde{u}|}. \tag{1.8}$$

Exercise 1.9 Check that if $m \geq 4$, div X has the same sign as $\tilde{u}$ in $\mathbb{R}^{2m}$.

Proof (of Theorem 1.8) By Exercise 1.9 we know that if $m \geq 4$, div X has the same sign as $\tilde{u}$, where $\tilde{u}$ and X are defined in (1.7) and (1.8). Let F be a competitor for E_S in a ball B_R, with F regular enough. We have that $F \setminus B_R = E_S \setminus B_R$.

Set $\Omega := F \setminus E_S$ (see Fig. 1.5). By using the fact that div $X \geq 0$ in Ω and the divergence theorem, we deduce that

$$0 \leq \int_\Omega \text{div } X \, dx = \int_{\partial E_S \cap \overline{\Omega}} X \cdot \nu_\Omega \, dH_{n-1} + \int_{\partial F \cap \overline{\Omega}} X \cdot \nu_\Omega \, dH_{n-1}. \tag{1.9}$$

Since $X = \nu_{E_S} = -\nu_\Omega$ on $\partial E_S \cap \overline{\Omega}$, and $|X| \leq 1$ (since in fact $|X| = 1$) everywhere (and hence in particular on $\partial F \cap \overline{\Omega}$), from (1.9) we conclude

$$H_{n-1}(\partial E_S \cap \overline{\Omega}) \leq H_{n-1}(\partial F \cap \overline{\Omega}). \tag{1.10}$$

With the same reasoning it is easy to prove that (1.10) holds also for $\Omega := E_S \setminus F$. Putting both inequalities together, we conclude that $P(E_S, B_R) \leq P(F, B_R)$.

Notice that the proof works for competitors F which are regular enough (since we applied the divergence theorem). However, it can be generalized to very

Fig. 1.5 A calibration
proving that the Simons cone
$\mathscr{C}_S$ is minimizing

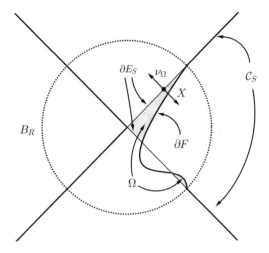

general competitors by using the generalized definition of perimeter, as in [23, Theorem 1.5]. □

Theorem 1.8 can also be proved with another argument—but still very much related to the previous one and that also uses the function $\tilde{u} = |x'|^4 - |x''|^4$. It consists of going to one more dimension $\mathbb{R}^{2m+1}$ and working with the minimal surface equation for graphs, (1.11) below. This is done in Theorem 16.4 of [28] (see also the proof of Theorem 2.2 in [13]).

In the proof above we used a vector field X satisfying the following three properties (with $E = E_S$):

(i)   div $X \geq 0$ in $B_R \setminus E$ and div $X \leq 0$ in $E \cap B_R$;
(ii) $X = \nu_E$ on $\partial E \cap B_R$;
(iii) $|X| \leq 1$ in B_R.

Definition 1.10 (Calibration)   If X satisfies the three properties above we say that X is a *calibration* for E in B_R.

Exercise 1.11   Use a similar argument to that of our last proof and build a calibration to show that a hyperplane in $\mathbb{R}^n$ is a minimizing minimal surface.

In an appendix, and with the purpose that the reader gets acquainted with another calibration, we present one which solves the isoperimetric problem: balls minimize perimeter among sets of given volume in $\mathbb{R}^n$. Note that the first variation (or Euler-Lagrange equation) for this problem is, by Lagrange multipliers, $\mathscr{H} = c$, where $c \in \mathbb{R}$ is a constant.

The following is an alternative proof of Theorem 1.8. It uses a *foliation argument*, as explained below. This second proof is probably more transparent (or intuitive) than the previous one and it is used often in minimal surfaces theory, but requires to know the existence of a (regular enough) minimizer (something that was not necessary in the previous proof). This existence result is available and can be proved with tools of the Calculus of Variations (see [17, 28]).

The proof also requires the use of the following important fact. If $\Sigma_1, \Sigma_2 \subset B_R$ are two connected hypersurfaces (regular enough), both satisfying $\mathcal{H} = 0$, and such that $\Sigma_1 \cap \Sigma_2 \neq \varnothing$ and Σ_1 lies on one side of Σ_2, then $\Sigma_1 \equiv \Sigma_2$ in B_R. Lying on one side can be defined as $\Sigma_1 = \partial F_1$, $\Sigma_2 = \partial F_2$, and $F_1 \subset F_2$. The same result holds if F_1 satisfies $\mathcal{H} = 0$ and F_2 satisfies $\mathcal{H} \geq 0$.

This result can be proved writing both surfaces as graphs in a neighborhood of a common point $P \in \Sigma_1 \cap \Sigma_2$. The minimal surface equation $\mathcal{H} = 0$ then becomes

$$\operatorname{div}\left(\frac{\nabla\varphi_1}{\sqrt{1 + |\nabla\varphi_1|^2}}\right) = 0 \tag{1.11}$$

for $\varphi_1 : \Omega \subset \mathbb{R}^{n-1} \to \mathbb{R}$ such that $\big(y', \varphi_1(y')\big) \subset \Omega \times \mathbb{R}$ is a piece of Σ_1 (after a rotation and translation). Then, assuming that φ_2 also satisfies (1.11)—or the appropriate inequality, one can see that $\varphi_1 - \varphi_2$ is a (super)solution of a second order linear elliptic equation. Since $\varphi_1 - \varphi_2 \geq 0$ (due to the ordering of Σ_1 and Σ_2), the strong maximum principle leads to $\varphi_1 - \varphi_2 \equiv 0$ (since $(\varphi_1 - \varphi_2)(0) = 0$ at the touching point). See Section 7 of Chapter 1 of [16] for more details.

Alternative proof (of Theorem 1.8) Note that the hypersurfaces

$$\left\{x \in \mathbb{R}^{2m} : \tilde{u}(x) = \lambda\right\},$$

with $\lambda \in \mathbb{R}$, form a foliation of $\mathbb{R}^{2m}$, where $\tilde{u}$ is the function defined in (1.7).

Let F be a minimizer of the perimeter in B_R among sets that coincide with E_S on ∂B_R, and assume that it is regular enough. Since F is a minimizer, in particular ∂F is a solution of the minimal surface equation $\mathcal{H} = 0$. Since $2m \geq 8$, by (1.6) and Exercise 1.9, the leaves of our foliation $\left\{x \in \mathbb{R}^{2m} : \tilde{u}(x) = \lambda\right\}$ are subsolutions of the same equation for $\lambda > 0$, and supersolutions for $\lambda < 0$.

If $F \not\equiv E_S$, there will be a first leaf (starting either from $\lambda = +\infty$ or from $\lambda = -\infty$) $\left\{x \in \mathbb{R}^{2m} : \tilde{u}(x) = \lambda_*\right\}$, with $\lambda_* \neq 0$, that touches ∂F at a point in $\overline{B}_R$ that we call P (see Fig. 1.6).

The point P cannot belong to ∂B_R, since it holds that

$$\partial F \cap \partial B_R = \mathscr{C}_S \cap \partial B_R = \{x : \tilde{u}(x) = 0\} \cap \partial B_R,$$

and the level sets of $\tilde{u}$ do not intersect each other. Thus, P must be an interior point of B_R. But then we arrive at a contradiction, by the "strong maximum principle" argument commented right before this proof, applied with $\Sigma_1 = \partial F$ and $\Sigma_2 = \left\{x \in \mathbb{R}^{2m} : \tilde{u}(x) = \lambda_*\right\}$.

As an exercise, write the details to prove the existence of a first leaf touching ∂F at an interior point.

Fig. 1.6 The foliation
argument to prove that the
Simons cone $\mathscr{C}_S$ is
minimizing

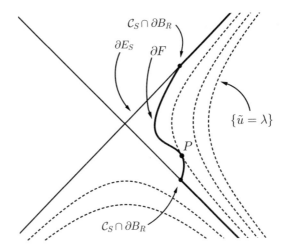

This same foliation argument will be used, in a simpler setting for graphs and the Allen-Cahn equation, in the proof of Theorem 1.32 in the next section.   □

Remark 1.12 The previous foliation argument gives more than the minimality of $\mathscr{C}_S$. It gives *uniqueness for the Dirichlet (or Plateau) problem* associated to the minimal surface equation with $\mathscr{C}_S$ as boundary value on ∂B_R.

Remark 1.13 In our alternative proof of Theorem 1.8 we used a clever foliation made of subsolutions and supersolutions. This sufficed to prove in a simple way Theorem 1.8, but required to (luckily) find the auxiliary function $\tilde{u} = |x'|^4 - |x''|^4$. Instead, in [5], Bombieri, De Giorgi, and Giusti considered the foliation made of exact solutions to the minimal surface equation $\mathscr{H} = 0$, when $2m \geq 8$. To this end, they proceeded as in the following exercise and wrote the minimal surface equation, for surfaces with rotational symmetry in x' and in x'', as an ODE in $\mathbb{R}^2$, finding Eq. (1.12) below. They then showed that the solutions of such ODE in the (s, t)-plane do not intersect each other (and neither the Simons cone), and thus form a foliation (see Remark 1.21 for more information on this).

Exercise 1.14 Let us set $s = |x'|$ and $t = |x''|$ for $x = (x', x'') \in \mathbb{R}^m \times \mathbb{R}^m$. Check that the following two ODEs are equivalent to the minimal surface equation $\mathscr{H} = 0$ written in the (s, t)-variables for surfaces with rotational symmetry in x' and in x''.

(i) As done in [5], if we set a parametric representation $s = s(\tau)$, $t = t(\tau)$, we find

$$s''t' - s't'' + (m - 1)\left((s')^2 + (t')^2\right)\left(\frac{s'}{t} - \frac{t'}{s}\right) = 0; \qquad (1.12)$$

(ii) as done in [19], if we set $s = e^{z(\theta)} \cos(\theta)$, $t = e^{z(\theta)} \sin(\theta)$ we get

$$z'' = \left(1 + (z')^2\right) \left((2m - 1) - \frac{2(m - 1) \cos(2\theta)}{\sin(2\theta)} z'\right).$$

The previous ODEs can be found starting from (1.6) when $u = u(s, t)$ depends only on s and t. Alternatively, they can also be found computing the first variation of the perimeter functional in $\mathbb{R}^{2m}$ written in the (s, t)-variables:

$$c \int s^{m-1} t^{m-1} \, dH_1(s, t), \tag{1.13}$$

for some positive constant c, that becomes

$$c \int e^{(2m-1)z(\theta)} \cos^{m-1}(\theta) \, \sin^{m-1}(\theta) \sqrt{1 + (z'(\theta))^2} \, d\theta$$

with the parametrization in point (ii).

Remark 1.15 For $n \geq 8$, there exist other minimizing cones, such as some of the *Lawson's cones*, defined by

$$\mathscr{C}_L = \left\{y = (y', y'') \in \mathbb{R}^k \times \mathbb{R}^{n-k} : |y'|^2 = c_{n,k} \, |y''|^2\right\} \quad \text{for } k \geq 2 \text{ and } n - k \geq 2.$$

For details, see [19].

Notice that if ∂E is a *cone* (i.e., $\lambda \partial E = \partial E$ for every $\lambda > 0$), in the expressions (1.1), (1.2), and (1.4) we will always consider ξ with compact support outside the origin (thus, not changing the possible singularity of the cone at the origin).

The next theorem was proved by Simons in 1968[1] (it is Theorem 10.10 in [28]). It is a crucial result towards the regularity theory of minimizing minimal surfaces.

Theorem 1.16 (Simons [36]) *Let $E \subset \mathbb{R}^n$ be an open set such that ∂E is a stable minimal cone and $\partial E \setminus \{0\}$ is regular. Thus, we are assuming $\mathscr{H} = 0$ and*

$$\int_{\partial E} \left\{|\delta\xi|^2 - c^2\xi^2\right\} dH_{n-1} \geq 0 \tag{1.14}$$

for every $\xi \in C^1(\partial E)$ with compact support outside the origin.
If $3 \leq n \leq 7$, then ∂E is a hyperplane.

---

[1]Theorem 1.16 was proved in 1965 by De Giorgi for $n = 3$, in 1966 by Almgren for $n = 4$, and finally in 1968 by Simons in any dimension $n \leq 7$.

Remark 1.17 Simons result (Theorem 1.16), together with a blow-up argument and a monotonicity formula (as main tools), lead to the flatness of every minimizing minimal surface in all of $\mathbb{R}^n$ if $n \leq 7$ (see [28, Theorem 17.3] for a proof). The same tools also give the analyticity of every minimal surface that is minimizing in a given ball of $\mathbb{R}^n$ if $n \leq 7$ (see [28, Theorem 10.11] for a detailed proof). See also [17] for a great shorter exposition of these results.

The dimension 7 in Theorem 1.16 is optimal, since by Theorem 1.8 the Simons cone provides a counterexample in dimension 8.

The following is a very rough explanation of why the minimizer of the Dirichlet (or Plateau) problem is the Simons cone (and thus passes through the origin) in high dimensions—in opposition with low dimensions, as in Fig. 1.4, where the minimizer stays away from the origin. In the perimeter functional written in the (s, t)-variables (1.13), the Jacobian $s^{m-1}t^{m-1}$ becomes smaller and smaller near the origin as m gets larger. Thus, lengths near $(s, t) = (0, 0)$ become smaller as the dimension m increases.

In order to prove Theorem 1.16, we start with some important preliminaries. Recalling (1.3), for $i = 1 \ldots, n$, we define the tangential derivative

$$\delta_i \xi := \partial_i \xi - \nu^i \, \nu^k \xi_k,$$

where $\nu = \nu_E = (\nu^1, \ldots, \nu^n) : \partial E \to \mathbb{R}^n$ is the exterior normal to E on ∂E, $\partial_i \xi = \partial_{x_i} \xi = \xi_i$ are Euclidean partial derivatives, and we used the standard convention of sum $\sum_{k=1}^{n}$ over repeated indices. As mentioned right after definition (1.3), even if to compute $\partial_i \xi$ requires to extend ξ to a neighborhood of ∂E, $\delta_i \xi$ is well defined knowing ξ only on ∂E—since it is a tangential derivative. Note also that we have n tangential derivatives $\delta_1, \ldots, \delta_n$ and, thus, they are linearly dependent, since ∂E is $(n - 1)$-dimensional. However, it is easy to check (as an exercise) that

$$|\delta \xi|^2 = \sum_{i=1}^{n} |\delta_i \xi|^2.$$

We next define *the Laplace-Beltrami operator* on ∂E by

$$\Delta_{LB} \xi := \sum_{i=1}^{n} \delta_i \delta_i \xi, \tag{1.15}$$

acting on functions $\xi : \partial E \to \mathbb{R}$. For the reader knowing Riemannian calculus, one can check that

$$\Delta_{LB} \xi = \mathrm{div}_T (\nabla_T \xi) = \mathrm{div}_T (\delta \xi),$$

where $\nabla_T = \delta$ is the tangential gradient introduced in (1.3) and div_T denotes the (tangential) divergence on the manifold ∂E.

According to (1.6), we have that

$$\mathscr{H} = \mathrm{div}_T \nu = \sum_{i=1}^{n} \delta_i \nu^i$$

We will also use the following *formula of integration by parts*:

$$\int_{\partial E} \delta_i \phi \, dH_{n-1} = \int_{\partial E} \mathscr{H} \phi \nu^i \, dH_{n-1} \tag{1.16}$$

for every (smooth) hypersurface ∂E and $\phi \in C^1(\partial E)$ with compact support. Equation (1.16) is proved in Giusti's book [28, Lemma 10.8]. However, there is a typo in the statement of [28, Lemma 10.8]: $\mathscr{H}$ is missed in the identity above, and there is an error of a sign in the proof of [28, Lemma 10.8].

Replacing ϕ by $\phi\varphi$ in (1.16), we deduce that

$$\int_{\partial E} \phi \, \delta_i \varphi \, dH_{n-1} = -\int_{\partial E} (\delta_i \phi)\varphi \, dH_{n-1} + \int_{\partial E} \mathscr{H} \phi \varphi \nu^i \, dH_{n-1}. \tag{1.17}$$

From this, replacing ϕ by $\delta_i \phi$ in (1.17) and using that $\sum_{i=1}^{n} \nu^i \delta_i \phi = \nu \cdot \delta\phi = 0$, we also have

$$\int_{\partial E} \delta\phi \cdot \delta\varphi \, dH_{n-1} = \sum_{i=1}^{n} \int_{\partial E} \delta_i \phi \, \delta_i \varphi \, dH_{n-1} = -\int_{\partial E} (\Delta_{LB} \phi)\varphi \, dH_{n-1}. \tag{1.18}$$

Remark 1.18 For a minimal surface ∂E, the second variation of perimeter given by (1.4) can also be rewritten, after (1.18), as

$$\int_{\partial E} \left\{ -\Delta_{LB} \xi - c^2 \xi \right\} \xi \, dH_{n-1}.$$

The operator $-\Delta_{LB} - c^2$ appearing in this expression is called *the Jacobi operator*. It is the linearization at the minimal surface ∂E of the minimal surface equation $\mathscr{H} = 0$.

Towards the proof of Simons theorem, let us now take $\xi = \tilde{c}\eta$ in (1.14), where $\tilde{c}$ and η are still arbitrary (η with compact support outside the origin) and will be chosen later. We obtain

$$0 \le \int_{\partial E} \left\{ |\delta\xi|^2 - c^2\xi^2 \right\} dH_{n-1}$$

$$= \int_{\partial E} \left\{ \tilde{c}^2|\delta\eta|^2 + \eta^2|\delta\tilde{c}|^2 + \tilde{c}\delta\tilde{c} \cdot \delta\eta^2 - c^2\tilde{c}^2\eta^2 \right\} dH_{n-1}$$

$$= \int_{\partial E} \left\{ \tilde{c}^2|\delta\eta|^2 - (\Delta_{LB}\tilde{c} + c^2\tilde{c})\tilde{c}\eta^2 \right\} dH_{n-1},$$

where at the last step we used integration by parts (1.17). This leads to the inequality

$$\int_{\partial E} \left\{ \Delta_{LB}\,\tilde{c} + c^2\tilde{c} \right\} \tilde{c}\eta^2 dH_{n-1} \le \int_{\partial E} \tilde{c}^2 |\delta\eta|^2 dH_{n-1},$$

where the term $\Delta_{LB}\,\tilde{c} + c^2\tilde{c}$ appearing in the first integral is the linearized or Jacobi operator at ∂E acting on $\tilde{c}$.

Now we make the choice $\tilde{c} = c$ and we arrive, as a consequence of stability, to

$$\int_{\partial E} \left\{ \frac{1}{2}\Delta_{LB}\,c^2 - |\delta c|^2 + c^4 \right\} \eta^2 dH_{n-1} \le \int_{\partial E} c^2 |\delta\eta|^2 dH_{n-1}. \tag{1.19}$$

At this point, Simons proof of Theorem 1.16 uses the following inequality for the Laplace-Beltrami operator Δ_{LB} of c^2 (recall that c^2 is the sum of the squares of the principal curvatures of ∂E), in the case when ∂E is a stationary cone.

Lemma 1.19 (Simons Lemma [36]) *Let $E \subset \mathbb{R}^n$ be an open set such that ∂E is a cone with zero mean curvature and $\partial E \setminus \{0\}$ is regular. Then, c^2 is homogeneous of degree -2 and, in $\partial E \setminus \{0\}$, we have*

$$\frac{1}{2}\Delta_{LB}\,c^2 - |\delta c|^2 + c^4 \ge \frac{2c^2}{|x|^2}.$$

In Sect. 1.1.3 we will give an outline of the proof of this result. We now use Lemma 1.19 to complete the proof of Theorem 1.16.

Proof (of Theorem 1.16) By using (1.19) together with Lemma 1.19 we obtain

$$0 \le \int_{\partial E} c^2 \left\{ |\delta\eta|^2 - \frac{2\eta^2}{|x|^2} \right\} dH_{n-1} \tag{1.20}$$

for every $\eta \in C^1(\partial E)$ with compact support outside the origin. By approximation, the same holds for η Lipschitz instead of C^1.

If $r = |x|$, we now choose η to be the Lipschitz function

$$\eta = \begin{cases} r^{-\alpha} & \text{if } r \le 1 \\ r^{-\beta} & \text{if } r \ge 1. \end{cases}$$

By directly computing

$$|\delta\eta|^2 = \begin{cases} \alpha^2 r^{-2\alpha-2} & \text{if } r \le 1 \\ \beta^2 r^{-2\beta-2} & \text{if } r \ge 1, \end{cases} \tag{1.21}$$

we realize that if

$$\alpha^2 < 2 \text{ and } \beta^2 < 2, \tag{1.22}$$

then in (1.20) we have $|\delta\eta|^2 - 2\eta^2/r^2 < 0$. If η were an admissible function in (1.20), we would then conclude that $c^2 \equiv 0$ on ∂E. This is equivalent to ∂E being an hyperplane.

Now, for η to have compact support and hence be admissible, we need to cut-off η near 0 and infinity. As an exercise, one can check that the cut-offs work (i.e., the tails in the integrals tend to zero) if (and only if)

$$\int_{\partial E} c^2 |\delta\eta|^2 dH_{n-1} < \infty, \tag{1.23}$$

or equivalently, since they have the same homogeneity,

$$\int_{\partial E} c^2 \frac{\eta^2}{|x|^2} dH_{n-1} < \infty.$$

By recalling that the Jacobian on ∂E (in spherical coordinates) is $r^{(n-1)-1}$, (1.21), and that, by Lemma 1.19, c^2 is homogeneous of degree -2, we deduce that (1.23) is satisfied if $n - 6 - 2\alpha > -1$ and $n - 6 - 2\beta < -1$. That is, if

$$\alpha < \frac{n-5}{2} \quad \text{and} \quad \frac{n-5}{2} < \beta. \tag{1.24}$$

If $3 \le n \le 7$ then $(n-5)^2/4 < 2$, i.e., $-\sqrt{2} < (n-5)/2 < \sqrt{2}$, and thus we can choose α and β satisfying (1.24) and (1.22). It then follows that $c^2 \equiv 0$, and hence ∂E is a hyperplane. $\qquad\qquad\qquad\qquad\qquad\qquad\qquad\qquad\qquad\qquad\qquad\qquad\qquad\square$

The argument in the previous proof (leading to the dimension $n \le 7$) is very much related to a well known result: Hardy's inequality in $\mathbb{R}^n$—which is presented next.

1.1.2 Hardy's Inequality

As already noticed in Remark 1.18, for a minimal surface ∂E the second variation of perimeter (1.4) can also be rewritten, by integrating by parts, as

$$\int_{\partial E} \left\{ -\Delta_{LB}\xi - c^2\xi \right\} \xi dH_{n-1}.$$

This involves the linearized or Jacobi operator $-\Delta_{LB} - c^2$. If $x = |x|\sigma = r\sigma$, with $\sigma \in S^{n-1}$, then $c^2 = d(\sigma)/|x|^2$ (if ∂E is a cone and thus c^2 is homogeneous of degree -2), where $d(\sigma)$ depends only on the angles σ. Thus, we are in the presence of the "Hardy-type operator"

$$-\Delta_{LB} - \frac{d(\sigma)}{|x|^2};$$

notice that Δ_{LB} and $d(\sigma)/|x|^2$ scale in the same way. Thus, for all admissible functions ξ,

$$0 \le \int_{\partial E} \left\{ |\delta \xi|^2 - \frac{d(\sigma)}{|x|^2} \xi^2 \right\} dH_{n-1}, \quad \text{if } \partial E \text{ is a stable minimal cone.}$$

Let us analyze the simplest case when $\partial E = \mathbb{R}^n$ and $d \equiv constant$. Then, the validity or not of the previous inequality is given by Hardy's inequality, stated and proved next.

Proposition 1.20 (Hardy's Inequality) *If $n \ge 3$ and $\xi \in C_c^1(\mathbb{R}^n \setminus \{0\})$, then*

$$\frac{(n-2)^2}{4} \int_{\mathbb{R}^n} \frac{\xi^2}{|x|^2} \, dx \le \int_{\mathbb{R}^n} |\nabla \xi|^2 \, dx. \tag{1.25}$$

In addition, $(n-2)^2/4$ is the best constant in this inequality and it is not achieved by any $0 \ne \xi \in H^1(\mathbb{R}^n)$.
  Moreover, if $a > (n-2)^2/4$, then the Dirichlet spectrum of $-\Delta_{LB} - a/|x|^2$ in the unit ball B_1 goes all the way to $-\infty$. That is,

$$\inf \frac{\int_{B_1} \{|\nabla \xi|^2 - a \frac{\xi^2}{|x|^2}\} dx}{\int_{B_1} |\xi|^2 dx} = -\infty, \tag{1.26}$$

where the infimum is taken over $0 \ne \xi \in H_0^1(B_1)$.

Proof Using spherical coordinates, for a given $\sigma \in S^{n-1}$ we can write

$$\int_0^{+\infty} r^{n-1} r^{-2} \xi^2(r\sigma) \, dr = -\frac{1}{n-2} \int_0^{+\infty} r^{n-2} 2\xi(r\sigma)\xi_r(r\sigma) \, dr. \tag{1.27}$$

Here we integrated by parts, using that $r^{n-3} = \left(r^{n-2}/(n-2)\right)'$.
  Now we apply the Cauchy-Schwarz inequality in the right-hand side to obtain

$$-\int_0^{+\infty} r^{n-2} \xi \xi_r \, r^{\frac{n-3}{2}} r^{-\frac{n-3}{2}} \, dr \le \left(\int_0^{+\infty} r^{n-3} \xi^2 \, dr\right)^{\frac{1}{2}} \left(\int_0^{+\infty} r^{n-1} \xi_r^2 \, dr\right)^{\frac{1}{2}}. \tag{1.28}$$

Putting together (1.27) and (1.28) we get

$$\int_0^{+\infty} r^{n-3} \xi^2 \, dr \le \frac{2}{n-2} \left(\int_0^{+\infty} r^{n-3} \xi^2 \, dr\right)^{\frac{1}{2}} \left(\int_0^{+\infty} r^{n-1} \xi_r^2 \, dr\right)^{\frac{1}{2}},$$

that is,

$$\frac{(n-2)^2}{4} \int_0^{+\infty} r^{n-1} \frac{\xi^2}{r^2} \, dr \leq \int_0^{+\infty} r^{n-1} \xi_r^2 \, dr.$$

By integrating in σ we conclude (1.25). An inspection of the equality cases in the previous proof shows that the best constant is not achieved.

Let us now consider $(n-2)^2/4 < \alpha^2 < a$ with $\alpha \searrow (n-2)/2$. Take

$$\xi = r^{-\alpha} - 1$$

and cut it off near the origin to be admissible. If we consider the main terms in the quotient (1.26), we get

$$\frac{\int (\alpha^2 - a) r^{-2\alpha - 2} dx}{\int r^{-2\alpha} dx}.$$

Thus it is clear that, as $\alpha \searrow (n-2)/2$, the denominator remains finite independently of the cut-off, while the numerator is as negative as we want after the cut-off. Hence, the quotient tends to $-\infty$.   □

Remark 1.21 As we explained in Remark 1.13, in [5], Bombieri, De Giorgi, and Giusti used a foliation made of exact solutions to the minimal surface equation $\mathscr{H} = 0$ when $2m \geq 8$. These are the solutions of the ODE (1.12) starting from points $(s(0), t(0)) = (s_0, 0)$ in the s-axis and with vertical derivative $(s'(0), t'(0)) = (0, 1)$. They showed that, for $2m \geq 8$, they do not intersect each other, neither intersect the Simons cone $\mathscr{C}_S$. Instead, in dimensions 4 and 6 they do not produce a foliation and, in fact, each of them crosses infinitely many times $\mathscr{C}_S$, as showed in Fig. 1.7. This reflects the fact that the linearized operator $-\Delta_{LB} - c^2$ on $\mathscr{C}_S$ has infinitely many negative eigenvalues, as in the simpler situation of Hardy's inequality in the last statement of Proposition 1.20.

1.1.3   Proof of the Simons Lemma

As promised, in this section we present the proof of Lemma 1.19 with almost all details. We follow the proof contained in Giusti's book [28], where more details can be found (Simons lemma is Lemma 10.9 in [28]). We point out that in the proof of [28] there are the following two typos:

- as already noticed before, the identity in the statement of [28, Lemma 10.8] is missing $\mathscr{H}$ in the second integrand. We wrote the corrected identity in Eq. (1.16) of these notes;
- the label (10.18) is missing in line -8, page 122 of [28].

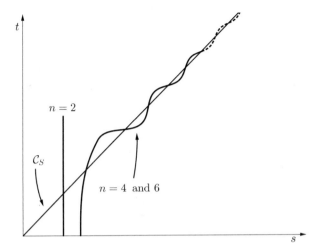

Fig. 1.7 Behaviour of the solutions to $\mathscr{H} = 0$ in dimensions 2, 4, and 6

Alternative proofs of Lemma 1.19 using intrinsic Riemaniann tensors can be found in the original paper of Simons [36] from 1968 and also in the book of Colding and Minicozzi [16].

Notation We denote by $d(x)$ the signed distance function to ∂E, defined by

$$d(x) := \begin{cases} \mathrm{dist}(x, \partial E), & x \in \mathbb{R}^n \setminus E, \\ -\,\mathrm{dist}(x, \partial E), & x \in E. \end{cases}$$

As we are assuming $E \setminus \{0\}$ to be regular, we have that $d(x)$ is C^2 in a neighborhood of $\partial E \setminus \{0\}$.

The normal vector to ∂E is given by

$$\nu = \nabla d = \frac{\nabla d}{|\nabla d|};$$

we write

$$\nu = (\nu^1, \ldots, \nu^n) = (d_1, \ldots, d_n),$$

where we adopt the abbreviated notation

$$w_i = w_{x_i} = \partial_i w \ \text{ and } \ w_{ij} = w_{x_i x_j} = \partial_{ij} w$$

for partial derivatives in $\mathbb{R}^n$. As introduced after Theorem 1.16, we will use the tangential derivatives

$$\delta_i := \partial_i - \nu^i \nu^k \partial_k$$

for $i = 1, \ldots, n$, and thus

$$\delta_i w = w_i - v^i v^k w_k,$$

where we adopted the summation convention over repeated indices. Finally, recall the Laplace-Beltrami operator defined in (1.15):

$$\Delta_{LB} := \delta_i \delta_i.$$

Remark 1.22 Since

$$1 = |v|^2 = \sum_{k=1}^{n} d_k^2, \tag{1.29}$$

it holds that

$$d_{jk} d_k = 0 \text{ for } j = 1, \ldots, n.$$

Thus, we have

$$\delta_i v^j = \delta_i d_j = d_{ij} - d_i d_k d_{kj} = d_{ij} = d_{ji},$$

which leads to

$$\delta_i v^j = \delta_j v^i.$$

Exercise 1.23 Using $\delta_i v^j = d_{ij}$, verify that

$$\mathcal{H} = \delta_i v^i,$$

$$c^2 = \delta_i v^j \delta_j v^i = \sum_{i,j=1}^{n} (\delta_i v^j)^2. \tag{1.30}$$

The identities

$$v^i \delta_i = 0,$$

$$v^i \delta_j v^i = 0, \text{ for } j = 1, \ldots, n, \tag{1.31}$$

will be used often in the following computations. The first one follows from the definition of δ_i, while the second is immediate from (1.29).

The next lemma will be useful in what follows.

Lemma 1.24 *The following equations hold for every smooth hypersurface ∂E:*

$$\delta_i \delta_j = \delta_j \delta_i + (v^i \delta_j v^k - v^j \delta_i v^k) \delta_k, \tag{1.32}$$

$$\Delta_{LB} v^j + c^2 v^j = \delta_j \mathcal{H} \ (= 0 \text{ if } \partial E \text{ is stationary}), \tag{1.33}$$

for all indices i and j.

For a proof of this lemma, see [28, Lemma 10.7].

Equation (1.33) is an important one. It says that the normal vector v to a minimal surface solves the Jacobi equation $(\Delta_{LB} + c^2) v \equiv 0$ on ∂E. This reflects the invariance of the equation $\mathcal{H} = 0$ by translations (to see this, write a perturbation made by a small translation as a normal deformation, as in Fig. 1.2).

If ∂E is stationary, from (1.32) and by means of simple calculations, one obtains that

$$\Delta_{LB} \delta_k = \delta_k \Delta_{LB} - 2v^k (\delta_i v^j) \delta_i \delta_j - 2(\delta_k v^j)(\delta_j v^i) \delta_i. \tag{1.34}$$

Equation (1.34) is the formula with the missed label (10.18) in [28].

We are ready now to give the

Outline of the proof (of Lemma 1.19) By (1.30) we can write that

$$\frac{1}{2} \Delta_{LB} c^2 = (\delta_i v^j) \Delta_{LB} \delta_i v^j + \sum_{i,j,k} (\delta_k \delta_i v^j)^2.$$

Then, using (1.33), (1.34), and the fact $\mathcal{H} = 0$, we have

$$\frac{1}{2} \Delta_{LB} c^2 = -(\delta_i v^j) \delta_i (c^2 v^j) - 2(\delta_i v^j)(\delta_k v^l)(\delta_l v^j)(\delta_i v^k) + \sum_{i,j,k} (\delta_k \delta_i v^j)^2,$$

and by (1.32)

$$\frac{1}{2} \Delta_{LB} c^2 = -c^4 - 2v^i v^l (\delta_j \delta_l v^k)(\delta_k \delta_i v^j) + \sum_{i,j,k} (\delta_k \delta_i v^j)^2.$$

Now, if $x_0 \in \partial E \setminus \{0\}$, we can choose the x_n-axis to be the same direction as $v(x_0)$. Thus, $v(x_0) = (0, \ldots, 0, 1)$ and at x_0 we have

$$v^n = 1, \ \delta_n = 0,$$

$$v^\alpha = 0, \ \delta_\alpha = \partial_\alpha \text{ for } \alpha = 1, \ldots, n-1.$$

Hence, computing from now on always at the point x_0, and by using (1.32) and (1.31), we get

$$\frac{1}{2}\Delta_{LB}\,c^2 = -c^4 + \sum_{\alpha,\beta,\gamma}(\delta_\gamma\delta_\alpha v^\beta)^2 + 2\sum_{\alpha,\gamma}(\delta_\gamma\delta_\alpha v^n)^2 - 2\sum_{\alpha,\beta}(\delta_\alpha\delta_\beta v^n)^2$$

$$= -c^4 + \sum_{\alpha,\beta,\gamma}(\delta_\gamma\delta_\alpha v^\beta)^2,$$

where all the greek indices indicate summation from 1 to $n-1$.

On the other hand, we have

$$|\delta c|^2 = \frac{1}{c^2}(\delta_\alpha v^\beta)(\delta_\gamma\delta_\alpha v^\beta)(\delta_\sigma v^\tau)(\delta_\gamma\delta_\sigma v^\tau),$$

and hence

$$\frac{1}{2}\Delta_{LB}\,c^2 + c^4 - |\delta c|^2 = \frac{1}{2c^2}\sum_{\alpha,\beta,\gamma,\sigma,\tau}\left[(\delta_\sigma v^\tau)(\delta_\gamma\delta_\alpha v^\beta) - (\delta_\alpha v^\beta)(\delta_\gamma\delta_\sigma v^\tau)\right]^2.$$

Now remember that ∂E is a cone with vertex at the origin, and thus $<x,v>=0$ on ∂E. Since we took $v=(0,\ldots,0,1)$ at x_0, we may choose coordinates in such a way that x_0 lies on the $(n-1)$-axis. In particular, $v^{n-1}=0$ at x_0 and

$$0 = \delta_i <x,v>=<\delta_i x,v> + <x,\delta_i v>=<x,\delta_i v>,$$

which leads to

$$\delta_i v^{n-1} = 0 \text{ at } x_0.$$

If now the letters A,B,S,T run from 1 to $n-2$, he have

$$\frac{1}{2}\Delta_{LB}\,c^2 + c^4 - |\delta c|^2 = \frac{1}{2c^2}\sum_{A,B,S,T,\gamma}\left[(\delta_S v^T)(\delta_\gamma\delta_A v^B) - (\delta_A v^B)(\delta_\gamma\delta_S v^T)\right]^2$$

$$+ \frac{2}{c^2}\sum_{S,T,\gamma,\alpha}(\delta_S v^T)^2(\delta_\gamma\delta_{n-1}v^\alpha)^2 \geq 2\sum_{\alpha,\gamma}(\delta_\gamma\delta_{n-1}v^\alpha)^2.$$

From (1.32), $\delta_i\delta_{n-1} = \delta_{n-1}\delta_i$ and $\delta_{n-1} = \partial_{n-1} = \pm\left(x^j/|x|\right)\partial_j$ at x_0. Since ∂E is a cone, v is homogeneous of degree 0 and hence $\delta_i v^\alpha$ is homogeneous of degree -1. Thus, by Euler's theorem on homogeneous functions, we have

$$\delta_{n-1}\delta_i v^\alpha = \pm\frac{x^j\partial_j}{|x|}\delta_i v^\alpha = \mp\frac{1}{|x|}\delta_i v^\alpha,$$

and hence

$$2 \sum_{i,\alpha} (\delta_i \delta_{n-1} v^\alpha)^2 = \frac{2}{|x|^2} \sum_{i,\alpha} (\delta_i v^\alpha)^2 = \frac{2c^2}{|x|^2}.$$

The proof is now completed. $\qquad\qquad\qquad\qquad\qquad\qquad\qquad\qquad\qquad$ □

1.1.4 Comments On: Harmonic Maps, Free Boundary Problems, and Nonlocal Minimal Surfaces

Here we briefly sketch arguments and results similar to the previous ones on minimal surfaces, now for three other elliptic problems.

1.1.4.1 Harmonic Maps

Consider the energy

$$E(u) = \frac{1}{2} \int_\Omega |Du|^2 dx \tag{1.35}$$

for H^1 maps $u : \Omega \subset \mathbb{R}^n \to \overline{S^N_+}$ from a domain Ω of $\mathbb{R}^n$ into the closed upper hemisphere

$$\overline{S^N_+} = \{y \in \mathbb{R}^{N+1} : |y| = 1, \ y_{N+1} \geq 0\}.$$

A critical point of E is called a (weakly) *harmonic map*. When a map minimizes E among all maps with values into $\overline{S^N_+}$ and with same boundary values on $\partial\Omega$, then it is called a minimizing harmonic map.

From the energy (1.35) and the restriction $|u| \equiv 1$, one finds that the equation for harmonic maps is given by

$$-\Delta u = |Du|^2 u \qquad \text{in } \Omega.$$

In 1983, Jäger and Kaul proved the following theorem, that we state here without proving it (see the original paper [29] for the proof).

Theorem 1.25 (Jäger-Kaul [29]) *The equator map*

$$u_* : B_1 \subset \mathbb{R}^n \to \overline{S^n_+}, \qquad x \mapsto (x/|x|, 0)$$

is a minimizing harmonic map on the class

$$\mathscr{C} = \{u \in H^1(B_1 \subset \mathbb{R}^n, S^n) : u = u_* \text{ on } \partial B_1\}$$

if and only if $n \geq 7$.

We just mention that the proof of the "if" in Theorem 1.25 uses a calibration argument.

Later, Giaquinta and Souček [27], and independently Schoen and Uhlenbeck [35], proved the following result.

Theorem 1.26 (Giaquinta-Souček [27]; Schoen-Uhlenbeck [35])

Let $u : B_1 \subset \mathbb{R}^n \to \overline{S_+^N}$ be a minimizing harmonic map, homogeneous of degree zero, into the closed upper hemisphere $\overline{S_+^N}$. If $3 \leq n \leq 6$, then u is constant.

Now we will show an outline of the proof of Theorem 1.26 following [27]. More details can also be found in Section 3 of [13]. This theorem gives an alternative proof of one part of the statement of Theorem 1.25. Namely, that the equator map u_* is not minimizing for $3 \leq n \leq 6$.

Outline of the proof (of Theorem 1.26) After stereographic projection (with respect to the south pole) P from $S^N \subset \mathbb{R}^{N+1}$ to $\mathbb{R}^N$, for the new function $v = P \circ u : B_1 \subset \mathbb{R}^n \to \mathbb{R}^N$, the energy (1.35) (up to a constant factor) is given by

$$E(v) := \int_{B_1} \frac{|Dv|^2}{(1 + |v|^2)^2} dx.$$

In addition, we have $|v| \leq 1$ since the image of u is contained in the closed upper hemisphere.

By testing the function

$$\xi(x) = v(x)\eta(|x|),$$

where η is a smooth radial function with compact support in B_1, in the equation of the first variation of the energy, that is

$$\delta E(v)\xi = 0,$$

one can deduce that either v is constant (and then the proof is finished) or

$$|v| \equiv 1,$$

that we assume from now on.

Since v is a minimizer, we have that the second variation of the energy satisfies

$$\delta^2 E(v)(\xi, \xi) \geq 0.$$

By choosing here the function

$$\xi(x) = v(x)|Dv(x)|\eta(|x|),$$

where η is a smooth radial function with compact support in B_1 (to be chosen later), and setting

$$c(x) := |Dv(x)|,$$

one can conclude the proof by similar arguments as in the previous section and by using Lemma 1.27, stated next. □

Lemma 1.27 *If v is a harmonic map, homogeneous of degree zero, and with $|v| \equiv 1$, we have*

$$\frac{1}{2}\Delta c^2 - |Dc|^2 + c^4 \geq \frac{c^2}{|x|^2} + \frac{c^4}{n-1},$$

where $c := |Dv|$.

This lemma is the analogue result of Lemma 1.19 for minimal cones. See [27] for a proof of the lemma, which also follows from Bochner identity (see [35]).

1.1.4.2 Free Boundary Problems

Consider the one-phase free boundary problem:

$$\begin{cases} \Delta u = 0 & \text{in } E \\ u = 0 & \text{on } \partial E \\ |\nabla u| = 1 & \text{on } \partial E \setminus \{0\}, \end{cases} \tag{1.36}$$

where u is homogeneous of degree one and positive in the domain $E \subset \mathbb{R}^n$ and ∂E is a cone. We are interested in solutions u that are stable for the Alt-Caffarelli energy functional

$$E_{B_1}(u) = \int_{B_1} \left\{|\nabla u|^2 + \chi_{\{u>0\}}\right\} dx$$

with respect to compact domain deformations that do not contain the origin. More precisely, we say that u is *stable* if for any smooth vector field $\Psi : \mathbb{R}^n \to \mathbb{R}^n$ with $0 \notin \text{supp}\Psi \subset B_1$ we have

$$\frac{d^2}{dt^2} E_{B_1}\left(u\left(x + t\Psi(x)\right)\right)\bigg|_{t=0} \geq 0.$$

The following result due to Jerison and Savin is contained in [30], where a detailed proof can be found.

Theorem 1.28 (Jerison-Savin [30]) *The only stable, homogeneous of degree one, solutions of (1.36) in dimension $n \leq 4$ are the one-dimensional solutions $u = (x \cdot v)^+$, $v \in S^{n-1}$.*

In dimension $n = 3$ this result had been established by Caffarelli et al. [14], where they conjectured that it remains true up to dimension $n \leq 6$. On the other hand, in dimension $n = 7$, De Silva and Jerison [22] provided an example of a nontrivial minimizer.

The proof of Jerison and Savin of Theorem 1.28 is similar to Simons proof of the rigidity of stable minimal cones in low dimensions: they find functions c (now involving the second derivatives of u) which satisfy appropriate differential inequalities for the linearized equation.

Here, the linearized problem is the following:

$$\begin{cases} \Delta v = 0 & \text{in } E \\ v_\nu + \mathscr{H} v = 0 & \text{on } \partial E \setminus \{0\} . \end{cases}$$

For the function

$$c^2 = \|D^2 u\|^2 = \sum_{i,j=1}^{n} u_{ij}^2,$$

they found the following interior inequality which is similar to the one of the Simons lemma:

$$\frac{1}{2} \Delta c^2 - |\nabla c|^2 \geq 2 \frac{n-2}{n-1} \frac{c^2}{|x|^2} + \frac{2}{n-1} |\nabla c|^2.$$

In addition, they also need to prove a boundary inequality involving c_ν. Furthermore, to establish Theorem 1.28 in dimension $n = 4$, a more involved function c of the second derivatives of u is needed.

1.1.4.3 Nonlocal Minimal Surfaces

Nonlocal minimal surfaces, or α-minimal surfaces (where $\alpha \in (0, 1)$), have been introduced in 2010 in the seminal paper of Caffarelli, Roquejoffre, and Savin [15]. These surfaces are connected to fractional perimeters and diffusions and, as $\alpha \nearrow 1$, they converge to classical minimal surfaces. We refer to the lecture notes [17] and the survey [24], where more references can be found.

For α-minimal surfaces and all $\alpha \in (0, 1)$, the analogue of Simons flatness result is only known in dimension 2 by a result for minimizers of Savin and Valdinoci [34].

1.2   The Allen-Cahn Equation

This section concerns the *Allen-Cahn equation*

$$- \Delta u = u - u^3 \quad \text{in } \mathbb{R}^n. \tag{1.37}$$

By using Eq. (1.37) and the maximum principle it can be proved that any solution satisfies $|u| \le 1$. Then, by the strong maximum principle we have that either $|u| < 1$ or $u \equiv \pm 1$. Since $u \equiv \pm 1$ are trivial solutions, from now on we consider $u : \mathbb{R}^n \to (-1, 1)$.

We introduce the class of *1-d solutions*:

$$u(x) = u_*(x \cdot e) \quad \text{for a vector } e \in \mathbb{R}^n, |e| = 1,$$

where

$$u_*(y) = \tanh\left(\frac{y}{\sqrt{2}} \right).$$

The solution u_* is sometimes referred to as the *layer solution* to (1.37); see Fig. 1.8. The fact that u depends only on one variable can be rephrased also by saying that all the level sets $\{u = s\}$ of u are hyperplanes.

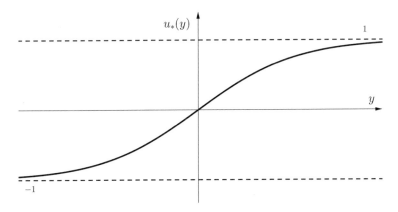

Fig. 1.8   The increasing, or layer, solution to the Allen-Cahn equation

Fig. 1.9 The double-well
potential in the Allen-Cahn
energy

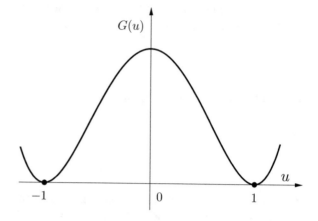

Exercise 1.29 Check that the 1-d functions introduced above are solutions of the Allen-Cahn equation.

Remark 1.30 Let us take $e = e_n = (0, \ldots, 0, 1)$ and consider the 1-d solution $u(x) = u_*(x_n) = \tanh(x_n/\sqrt{2})$. It is clear that the following two relations hold:

$$u_{x_n} > 0 \quad \text{in } \mathbb{R}^n, \tag{1.38}$$

$$\lim_{x_n \to \pm\infty} u(x', x_n) = \pm 1 \quad \text{for all } x' \in \mathbb{R}^{n-1}. \tag{1.39}$$

The energy functional associated to Eq. (1.37) is

$$E_\Omega(u) := \int_\Omega \left\{ \frac{1}{2} |\nabla u|^2 + G(u) \right\} dx,$$

where G is the double-well potential in Fig. 1.9:

$$G(u) = \frac{1}{4} \left(1 - u^2 \right)^2.$$

Definition 1.31 (Minimizer) A function $u : \mathbb{R}^n \to (-1, 1)$ is said to be a *minimizer* of (1.37) when

$$E_{B_R}(u) \leq E_{B_R}(v)$$

for every open ball B_R and functions $v : \overline{B_R} \to \mathbb{R}$ such that $v \equiv u$ on ∂B_R.

Connection with the Theory of Minimal Surfaces The Allen-Cahn equation has its origin in the theory of phase transitions and it is used as a model for some nonlinear reaction-diffusion processes. To better understand this, let $\Omega \subset \mathbb{R}^n$ be a bounded domain, and consider the Allen-Cahn equation with parameter $\varepsilon > 0$,

$$- \varepsilon^2 \Delta u = u - u^3 \ \text{in} \ \Omega, \tag{1.40}$$

with associated energy functional given by

$$E_\varepsilon(u) = \int_\Omega \left\{ \frac{\varepsilon}{2} |\nabla u|^2 + \frac{1}{\varepsilon} G(u) \right\} dx. \tag{1.41}$$

Assume now that there are two populations (or chemical states) A and B and that u is a density measuring the percentage of the two populations at every point: if $u(x) = 1$ (respectively, $u(x) = -1$) at a point x, we have only population A at x (respectively, population B); $u(x) = 0$ means that at x we have 50% of population A and 50% of population B.

By (1.41), it is clear that in order to minimize E_ε as ε tends to 0, $G(u)$ must be very small. From Fig. 1.9 we see that this happens when u is close to ± 1. These heuristics are indeed formally confirmed by a celebrated theorem of Modica and Mortola. It states that, if u_ε is a family of minimizers of E_ε, then, up to a subsequence, u_ε converges in $L^1_{\text{loc}}(\Omega)$, as ε tends to 0, to

$$u_0 = \chi_{\Omega_+} - \chi_{\Omega_-}$$

for some disjoint sets $\Omega_\pm$ having as common boundary a surface Γ. In addition, Γ is a minimizing minimal surface. Therefore, the result of Modica-Mortola establishes that the two populations tend to their total separation, and in such a (clever) way that the interface surface Γ of separation has least possible area.

Finally, notice that the 1-d solution of (1.40),

$$x \mapsto u_*\left(\frac{x \cdot e}{\varepsilon}\right),$$

makes a very fast transition from -1 to 1 in a scale of order ε. Accordingly, in Fig. 1.10, u_ε will make this type of fast transition across the limiting minimizing minimal surface Γ. The interested reader can see [1] for more details.

1.2.1 Minimality of Monotone Solutions with Limits ± 1

The following fundamental result shows that monotone solutions with limits ± 1 are minimizers (as in Definition 1.31).

Fig. 1.10 The zero level set of u_ε, the limiting function u_0, and the minimal surface Γ

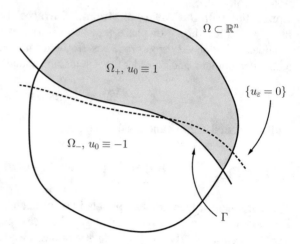

$\Omega \subset \mathbb{R}^n$

$\Omega_+, u_0 \equiv 1$

$\{u_\varepsilon = 0\}$

$\Omega_-, u_0 \equiv -1$

Γ

Theorem 1.32 (Alberti-Ambrosio-Cabré [1]) *Suppose that u is a solution of* (1.37) *satisfying the monotonicity hypothesis* (1.38) *and the condition* (1.39) *on limits. Then, u is a minimizer of* (1.37) *in $\mathbb{R}^n$.*

See [1] for the original proof of the Theorem 1.32. It uses a calibration built from a foliation and avoids the use of the strong maximum principle, but it is slightly involved. Instead, the simple proof that we give here was suggested to the first author (after one of his lectures on [1]) by L. Caffarelli. It uses a simple foliation argument together with the strong maximum principle, as in the alternative proof of Theorem 1.8 given in Sect. 1.1.1.

Proof (of Theorem 1.32) Denoting $x = (x', x_n) \in \mathbb{R}^{n-1} \times \mathbb{R}$, let us consider the functions

$$u^t(x) := u(x', x_n + t), \quad \text{for } t \in \mathbb{R}.$$

By the monotonicity assumption (1.38) we have that

$$u^t < u^{t'} \text{ in } \mathbb{R}^n, \text{ if } t < t'. \tag{1.42}$$

Thus, by (1.39) we have that the graphs of $u^t = u^t(x)$, $t \in \mathbb{R}$, form a foliation filling all of $\mathbb{R}^n \times (-1, 1)$. Moreover, we have that for every $t \in \mathbb{R}$, u^t are solutions of $-\Delta u^t = u^t - (u^t)^3$ in $\mathbb{R}^n$.

By simple arguments of the Calculus of Variations, given a ball B_R it can be proved that there exists a minimizer $v : \overline{B}_R \to \mathbb{R}$ of E_{B_R} such that $v = u$ on ∂B_R. In particular, v satisfies

$$\begin{cases} -\Delta v = v - v^3 & \text{in } B_R \\ |v| < 1 & \text{in } \overline{B}_R \\ v = u & \text{on } \partial B_R. \end{cases}$$

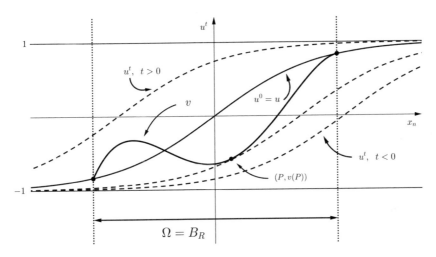

Fig. 1.11 The foliation $\{u^t\}$ and the minimizer v

By (1.39), we have that the graph of u^t in the compact set $\overline{B}_R$ is above the graph of v for t large enough, and it is below the graph of v for t negative enough (see Fig. 1.11). If $v \not\equiv u$, assume that $v < u$ at some point in B_R (the situation $v > u$ somewhere in B_R is done similarly). It follows that, starting from $t = -\infty$, there will exist a first $t_* < 0$ such that u^{t_*} touches v at a point $P \in \overline{B}_R$. This means that $u^{t_*} \leq v$ in $\overline{B}_R$ and $u^{t_*}(P) = v(P)$.

By (1.42), $t_* < 0$, and the fact that $v = u = u^0$ on ∂B_R, the point P cannot belong to ∂B_R. Thus, P will be an interior point of B_R.

But then we have that u^{t_*} and v are two solutions of the same semilinear equation (the Allen-Cahn equation), the graph of u^{t_*} stays below that of v, and they touch each other at the interior point $(P, v(P))$. This is a contradiction with the strong maximum principle.

Here we leave as an exercise (stated next) to verify that the difference of two solutions of $-\Delta u = f(u)$ satisfies a linear elliptic equation to which we can apply the strong maximum principle. This leads to $u^{t_*} \equiv v$, which contradicts $u^{t_*} < v = u^0$ on ∂B_R.   □

Exercise 1.33 Prove that the difference $w := v_1 - v_2$ of two solutions of a semilinear equation $-\Delta v = f(v)$, where f is a Lipschitz function, satisfies a linear equation of the form $\Delta w + c(x) w = 0$, for some function $c \in L^\infty$. Verify that, as a consequence, this leads to $u^{t_*} \equiv v$ in the previous proof.

By recalling Remark 1.30, we immediately get the following corollary.

Corollary 1.34 *The 1-d solution $u(x) = u_*(x \cdot e)$ is a minimizer of (1.37) in $\mathbb{R}^n$, for every unit vector $e \in \mathbb{R}^n$.*

As a corollary of Theorem 1.32, we easily deduce the following important energy estimates.

Corollary 1.35 (Energy Upper Bounds; Ambrosio-Cabré [2]) *Let u be a solution of (1.37) satisfying (1.38) and (1.39) (or more generally, let u be a minimizer in $\mathbb{R}^n$).*

Then, for all $R \geq 1$ we have

$$E_{B_R}(u) \leq C R^{n-1} \tag{1.43}$$

for some constant C independent of R. In particular, since $G \geq 0$, we have that

$$\int_{B_R} |\nabla u|^2 \, dx \leq C R^{n-1}$$

for all $R \geq 1$.

Remark 1.36 The proof of Corollary 1.35 is trivial for 1-d solutions. Indeed, it is easy to check that $\int_{-\infty}^{+\infty} \left\{ \frac{1}{2}(u'_*)^2 + \frac{1}{4}(1 - u_*^2)^2 \right\} dy < \infty$ and, as a consequence, by applying Fubini's theorem on a cube larger than B_R, that (1.43) holds. This argument also shows that the exponent $n - 1$ in (1.43) is optimal (since it cannot be improved for 1-d solutions).

The estimates in Corollary 1.35 are fundamental in the proofs of a conjecture of De Giorgi that we treat in the next subsection.

The estimate (1.43) was first proved by Ambrosio and the first author in [2]. Later on, in [1] Alberti, Ambrosio, and the first author discovered that monotone solutions with limits are minimizers (Theorem 1.32 above). This allowed to simplify the original proof of the energy estimates found in [2], as follows.

Proof (of Corollary 1.35) Since u is a minimizer by Theorem 1.32 (or by hypothesis), we can perform a simple energy comparison argument. Indeed, let $\phi_R \in C^\infty(\mathbb{R}^n)$ satisfy $0 \leq \phi_R \leq 1$ in $\mathbb{R}^n$, $\phi_R \equiv 1$ in B_{R-1}, $\phi_R \equiv 0$ in $\mathbb{R}^n \setminus B_R$, and $\|\nabla \phi_R\|_\infty \leq 2$. Consider

$$v_R := (1 - \phi_R)u + \phi_R.$$

Since $v_R \equiv u$ on ∂B_R, we can compare the energy of u in B_R with that of v_R. We obtain

$$\int_{B_R} \left\{ \frac{1}{2}|\nabla u|^2 + G(u) \right\} dx \leq \int_{B_R} \left\{ \frac{1}{2}|\nabla v_R|^2 + G(v_R) \right\} dx$$

$$= \int_{B_R \setminus B_{R-1}} \left\{ \frac{1}{2}|\nabla v_R|^2 + G(v_R) \right\} dx \leq C|B_R \setminus B_{R-1}| \leq C R^{n-1}$$

for every $R \geq 1$, with C independent of R. In the second inequality of the chain above we used that $\frac{1}{2}|\nabla v_R|^2 + G(v_R) \leq C$ in $B_R \setminus B_{R-1}$ for some constant C independent of R. This is a consequence of the following exercise. $\qquad\qquad \square$

Exercise 1.37 Prove that if u is a solution of a semilinear equation $-\Delta u = f(u)$ in $\mathbb{R}^n$ and $|u| \leq 1$ in $\mathbb{R}^n$, where f is a continuous nonlinearity, then $|u| + |\nabla u| \leq C$ in $\mathbb{R}^n$ for some constant C depending only on n and f. See [2], if necessary, for a proof.

1.2.2 A Conjecture of De Giorgi

In 1978, E. De Giorgi [20] stated the following conjecture:

Conjecture (DG) Let $u : \mathbb{R}^n \to (-1, 1)$ be a solution of the Allen-Cahn equation (1.37) satisfying the monotonicity condition (1.38). Then, u is a 1-d solution (or equivalently, all level sets $\{u = s\}$ of u are hyperplanes), at least if $n \leq 8$.

This conjecture was proved in 1997 for $n = 2$ by Ghoussoub and Gui [26], and in 2000 for $n = 3$ by Ambrosio and Cabré [2]. Next we state a deep result of Savin [33] under the only assumption of minimality. This is the semilinear analogue of Simons Theorem 1.16 and Remark 1.17 on minimal surfaces. As we will see, Savin's result leads to a proof of Conjecture (DG) for $n \leq 8$ if the additional condition (1.39) on limits is assumed.

Theorem 1.38 (Savin [33]) Assume that $n \leq 7$ and that u is a minimizer of (1.37) in $\mathbb{R}^n$. Then, u is a 1-d solution.

The hypothesis $n \leq 7$ on its statement is sharp. Indeed, in 2017 Liu, Wang, and Wei [31] have shown the existence of a minimizer in $\mathbb{R}^8$ whose level sets are not hyperplanes. Its zero level set is asymptotic at infinity to the Simons cone. However, a canonical solution described in Sect. 1.2.3 (and whose zero level set is exactly the Simons cone) is still not known to be a minimizer in $\mathbb{R}^8$.

Note that Theorem 1.38 makes no assumptions on the monotonicity or the limits at infinity of the solution. To prove Conjecture (DG) using Savin's result (Theorem 1.38), one needs to make the further assumption (1.39) on the limits only to guarantee, by Theorem 1.32, that the solution is actually a minimizer. Then, Theorem 1.38 (and the gain of one more dimension, $n = 8$, thanks to the monotonicity of the solution) leads to the proof of Conjecture (DG) for monotone solutions with limits ± 1.

However, for $4 \leq n \leq 8$ the conjecture in its original statement (i.e., without the limits ± 1 as hypothesis) is still open. To our knowledge no clear evidence is known about its validity or not.

The proof of Theorem 1.38 uses an improvement of flatness result for the Allen-Cahn equation developed by Savin, as well as Theorem 1.16 on the non-existence of stable minimal cones in dimension $n \leq 7$.

Instead, the proofs of Conjecture (DG) in dimensions 2 and 3 are much simpler. They use the energy estimates of Corollary 1.35 and a Liouville-type theorem developed in [2] (see also [1]). As explained next, the idea of the proof originates in the paper [3] by Berestycki, Caffarelli, and Nirenberg.

Motivation for the Proof of Conjecture (DG) for $n \leq 3$ In [3] the authors made the following heuristic observation. From the equation $-\Delta u = f(u)$ and the monotonicity assumption (1.38), by differentiating we find that

$$u_{x_n} > 0 \text{ and } Lu_{x_n} := \left(-\Delta - f'(u)\right) u_{x_n} = 0 \text{ in } \mathbb{R}^n. \tag{1.44}$$

If we were in a bounded domain $\Omega \subset \mathbb{R}^n$ instead of $\mathbb{R}^n$ (and we forgot about boundary conditions), from (1.44), we would deduce that u_{x_n} is the first eigenfunction of L and that its first eigenvalue is 0. As a consequence, such eigenvalue is simple. But then, since we also have that

$$Lu_{x_i} = \left(-\Delta - f'(u)\right) u_{x_i} = 0 \quad \text{for } i = 1, \ldots, n-1,$$

the simplicity of the eigenvalue would lead to

$$u_{x_i} = c_i u_{x_n} \quad \text{for } i = 1, \ldots, n-1, \tag{1.45}$$

where c_i are constants. Now, we would conclude that u is a 1-d solution, by the following exercise.

Exercise 1.39 Check that (1.45), with c_i being constants, is equivalent to the fact that u is a 1-d solution.

To make this argument work in the whole $\mathbb{R}^n$, one needs a Liouville-type theorem. For $n = 2$ it was proved in [3] and [26]. Later, [2] used it to prove Conjecture (DG) in $\mathbb{R}^3$ after proving the crucial energy estimate (1.43). The Liouville theorem requires the right hand side of (1.43) to be bounded by $CR^2 = CR^{3-1}$.

In 2011, del Pino, Kowalczyk, and Wei [21] established that Conjecture (DG) does not hold for $n \geq 9$—as suggested in De Giorgi's original statement.

Theorem 1.40 (del Pino-Kowalczyk-Wei [21]) *If $n \geq 9$, there exists a solution of* (1.37), *satisfying* (1.38) *and* (1.39), *and which is not a 1-d solution.*

The proof in [21] uses crucially the minimal graph in $\mathbb{R}^9$ built by Bombieri, De Giorgi, and Giusti in [5]. This is a minimal surface in $\mathbb{R}^9$ given by the graph of a function $\phi : \mathbb{R}^8 \to \mathbb{R}$ which is antisymmetric with respect to the Simons cone. The solution of Theorem 1.40 is built in such a way that its zero level set stays at finite distance from the Bombieri-De Giorgi-Giusti graph.

We consider next a similar object to the previous minimal graph, but in the context of the Allen-Cahn equation: a solution $u : (\mathbb{R}^8 =)\mathbb{R}^{2m} \to \mathbb{R}$ which is antisymmetric with respect to the Simons cone.

1.2.3   The Saddle-Shaped Solution Vanishing on the Simons Cone

As in Sect. 1.1, let $m \geq 1$ and denote by $\mathscr{C}_S$ the Simons cone (1.5). For $x = (x_1, \ldots, x_{2m}) \in \mathbb{R}^{2m}$, s and t denote the two radial variables

$$s = \sqrt{x_1^2 + \ldots + x_m^2} \quad \text{and} \quad t = \sqrt{x_{m+1}^2 + \ldots + x_{2m}^2}. \tag{1.46}$$

The Simons cone is given by

$$\mathscr{C}_S = \{s = t\} = \partial E, \quad \text{where } E = \{s > t\}.$$

Definition 1.41 (Saddle-Shaped Solution)   We say that $u : \mathbb{R}^{2m} \to \mathbb{R}$ is a *saddle-shaped solution* (or simply a *saddle solution*) of the Allen-Cahn equation

$$- \Delta u = u - u^3 \quad \text{in } \mathbb{R}^{2m} \tag{1.47}$$

whenever u is a solution of (1.47) and, with s and t defined by (1.46),

(a) u depends only on the variables s and t. We write $u = u(s, t)$;
(b) $u > 0$ in $E := \{s > t\}$;
(c) $u(s, t) = -u(t, s)$ in $\mathbb{R}^{2m}$.

Remark 1.42 Notice that if u is a saddle-shaped solution, then we have $u = 0$ on $\mathscr{C}_S$ (see Fig. 1.12).

While the existence of a saddle-shaped solution is easily established, its uniqueness is more delicate. This was accomplished in 2012 by the first author in [9].

Fig. 1.12 The saddle-shaped solution u and the Simons cone $\mathscr{C}_S$

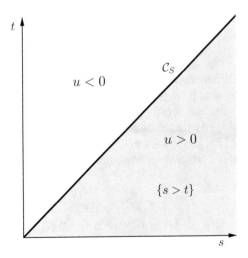

Theorem 1.43 (Cabré [9]) *For every even dimension* $2m \geq 2$, *there exists a unique saddle-shaped solution u of* (1.47).

Due to the minimality of the Simons cone when $2m \geq 8$ (and also because of the minimizer from [31] referred to after Theorem 1.38), the saddle-shaped solution is expected to be a minimizer when $2m \geq 8$:

Open problem 1.44 Is the saddle-shaped solution a minimizer of (1.47) in $\mathbb{R}^8$, or at least in higher even dimensions?

Nothing is known on this open problem except for the following result. It establishes stability (a weaker property than minimality) for $2m \geq 14$. Below, we sketch its proof.

Theorem 1.45 (Cabré [9]) *If* $2m \geq 14$, *the saddle-shaped solution u of* (1.47) *is stable in* $\mathbb{R}^{2m}$, *in the sense of the following definition.*

Definition 1.46 (Stability) We say that a solution u of $-\Delta u = f(u)$ in $\mathbb{R}^n$ is *stable* if the second variation of the energy with respect to compactly supported perturbations ξ is nonnegative. That is, if

$$\int_{\mathbb{R}^n} \left\{ |\nabla \xi|^2 - f'(u)\xi^2 \right\} dx \geq 0 \quad \text{for all } \xi \in C_c^1(\mathbb{R}^n).$$

In the rest of this section, we will take $n = 2m$ and f to be the Allen-Cahn nonlinearity, i.e., $f(u) = u - u^3$.

Outline of the proof (of Theorem 1.45) Notice that

$$u_{ss} + u_{tt} + (m - 1)\left(\frac{u_s}{s} + \frac{u_t}{t}\right) + f(u) = 0, \tag{1.48}$$

for $s > 0$ and $t > 0$, is equation (1.47) expressed in the (s, t) variables. Let us introduce the function

$$\varphi := t^{-b} u_s - s^{-b} u_t. \tag{1.49}$$

Differentiating (1.48) with respect to s (and to t), one finds equations satisfied by u_s (and by u_t)—and which involve a zero order term with coefficient $f'(u)$. These equations, together with some more delicate monotonicity properties of the saddle-shaped solution established in [9], can be used to prove the following fact.

For $2m \geq 14$, one can choose $b > 0$ in (1.49) (see [9] for more details) such that φ is a positive supersolution of the linearized problem, i.e.:

$$\varphi > 0 \quad \text{in } \{st > 0\}, \tag{1.50}$$

$$\{\Delta + f'(u)\}\varphi \leq 0 \quad \text{in } \mathbb{R}^{2m} \setminus \{st = 0\} = \{st > 0\}. \tag{1.51}$$

Next, using (1.50) and (1.51), we can verify the stability condition of u for any C^1 test function $\xi = \xi(x)$ with compact support in $\{st > 0\}$. Indeed, multiply (1.51) by ξ^2/φ and integrate by parts to get

$$
\int_{\{st>0\}} f'(u)\,\xi^2\,dx = \int_{\{st>0\}} f'(u)\varphi\,\frac{\xi^2}{\varphi}\,dx
$$

$$
\leq \int_{\{st>0\}} -\Delta\varphi\,\frac{\xi^2}{\varphi}\,dx
$$

$$
= \int_{\{st>0\}} \nabla\varphi\,\nabla\xi\,\frac{2\xi}{\varphi}\,dx - \int_{\{st>0\}} \frac{|\nabla\varphi|^2}{\varphi^2}\,\xi^2\,dx.
$$

Now, using the Cauchy-Schwarz inequality, we are led to

$$
\int_{\{st>0\}} f'(u)\,\xi^2\,dx \leq \int_{\{st>0\}} |\nabla\xi|^2\,dx.
$$

Finally, by a cut-off argument we can prove that this same inequality holds also for every function $\xi \in C_c^1(\mathbb{R}^{2m})$. $\square$

Remark 1.47 Alternatively to the variational proof seen above, another way to establish stability from the existence of a positive supersolution to the linearized problem is by using the maximum principle (see [4] for more details).

1.3 Blow-Up Problems

In this final section, we consider positive solutions of the semilinear problem

$$
\begin{cases}
-\Delta u = f(u) & \text{in } \Omega \\
u > 0 & \text{in } \Omega \\
u = 0 & \text{on } \partial\Omega,
\end{cases}
\tag{1.52}
$$

where $\Omega \subset \mathbb{R}^n$ is a smooth bounded domain, $n \geq 1$, and $f : \mathbb{R}^+ \to \mathbb{R}$ is C^1.

The associated energy functional is

$$
E_\Omega(u) := \int_\Omega \left\{ \frac{1}{2}|\nabla u|^2 - F(u) \right\} dx,
\tag{1.53}
$$

where F is such that $F' = f$.

1.3.1 Stable and Extremal Solutions: A Singular Stable Solution for $n \geq 10$

We define next the class of stable solutions to (1.52). It includes any local minimizer, i.e., any minimizer of (1.53) under small perturbations vanishing on $\partial\Omega$.

Definition 1.48 (Stability) A solution u of (1.52) is said to be *stable* if the second variation of the energy with respect to C^1 perturbations ξ vanishing on $\partial\Omega$ is nonnegative. That is, if

$$\int_\Omega f'(u)\xi^2\,dx \leq \int_\Omega |\nabla\xi|^2\,dx \quad \text{for all } \xi \in C^1(\overline{\Omega}) \text{ with } \xi_{|\partial\Omega} \equiv 0. \tag{1.54}$$

There are many nonlinearities for which (1.52) admits a (positive) stable solution. Indeed, replace $f(u)$ by $\lambda f(u)$ in (1.52), with $\lambda \geq 0$:

$$\begin{cases} -\Delta u = \lambda f(u) & \text{in } \Omega \\ \quad u = 0 & \text{on } \partial\Omega. \end{cases} \tag{1.55}$$

Assume that f is positive, nondecreasing, and superlinear at $+\infty$, that is,

$$f(0) > 0, \quad f' \geq 0 \quad \text{and} \quad \lim_{t\to+\infty} \frac{f(t)}{t} = +\infty. \tag{1.56}$$

Note that also in this case we look for positive solutions (when $\lambda > 0$), since $f > 0$. We point out that, for $\lambda > 0$, $u \equiv 0$ is not a solution.

Proposition 1.49 *Assuming* (1.56), *there exists an extremal parameter $\lambda^* \in (0, +\infty)$ such that if $0 \leq \lambda < \lambda^*$ then* (1.55) *admits a minimal stable classical solution u_λ. Here "minimal" means the smallest among all the solutions, while "classical" means of class C^2. Being classical is a consequence of $u_\lambda \in L^\infty(\Omega)$ if $\lambda < \lambda^*$.*

On the other hand, if $\lambda > \lambda^*$ then (1.55) *has no classical solution.*

The family of classical solutions $\{u_\lambda : 0 \leq \lambda < \lambda^*\}$ is increasing in λ, and its limit as $\lambda \uparrow \lambda^*$ is a weak solution $u^* = u_{\lambda^*}$ of (1.55) *for $\lambda = \lambda^*$.*

Definition 1.50 (Extremal Solution) The function u^* given by Proposition 1.49 is called *the extremal solution* of (1.55).

For a proof of Proposition 1.49 see the book [25] by L. Dupaigne. The definition of weak solution (the sense in which u^* is a solution) requires $u^* \in L^1(\Omega)$, $f(u^*)\text{dist}(\cdot, \partial\Omega) \in L^1(\Omega)$, and the equation to be satisfied in the distributional sense after multiplying it by test functions vanishing on $\partial\Omega$ and integrating by parts twice (see [25]). Other useful references regarding extremal and stable solutions are [6, 7], and [11].

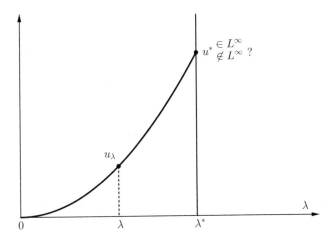

Fig. 1.13 The family of stable solutions u_λ and the extremal solution u^*

Since 1996, Brezis has raised several questions regarding stable and extremal solutions; see for instance [6]. They have led to interesting works, some of them described next. One of his questions is the following (see Fig. 1.13).

Question (Brezis) Depending on the dimension n or on the domain Ω, is the extremal solution u^* of (1.55) bounded (and therefore classical) or is it unbounded? More generally, one may ask the same question for the larger class of stable solutions to (1.52).

The following is an explicit example of stable unbounded (or singular) solution.
It is easy to check that, for $n \geq 3$, the function $\tilde{u} = -2 \log |x|$ is a solution of (1.52) in $\Omega = B_1$, the unit ball, for $f(u) = 2(n-2)e^u$. Let us now consider the linearized operator at $\tilde{u}$, which is given by

$$-\Delta - 2(n-2)e^{\tilde{u}} = -\Delta - \frac{2(n-2)}{|x|^2}.$$

If $n \geq 10$, then its first Dirichlet eigenvalue in B_1 is nonnegative. This is a consequence of *Hardy's inequality* (1.25):

$$\frac{(n-2)^2}{4} \int_{B_1} \frac{\xi^2}{|x|^2} dx \ \leq \ \int_{B_1} |\nabla \xi|^2 dx \qquad \text{for every } \xi \in H_0^1(B_1),$$

and the fact that $2(n-2) \leq (n-2)^2/4$ if $n \geq 10$. Thus we proved the following result.

Proposition 1.51 *For $n \geq 10$, $\tilde{u} = -2 \log |x|$ is an $H_0^1(B_1)$ stable weak solution of $-\Delta u = 2(n-2)e^u$ in B_1, $u > 0$ in B_1, $u = 0$ on ∂B_1.*

Thus, in dimensions $n \geq 10$ there exist unbounded H_0^1 stable weak solutions of (1.52), even in the unit ball and for the exponential nonlinearity. It is believed that $n \geq 10$ could be the optimal dimension for this fact, as we describe next.

1.3.2 Regularity of Stable Solutions. The Allard and Michael-Simon Sobolev Inequality

The following results give L^∞ bounds for stable solutions. To avoid technicalities we state the bounds for the extremal solution but, more generally, they also apply to every stable weak solution of (1.52) which is the pointwise limit of a sequence of bounded stable solutions to similar equations (see [25]).

Theorem 1.52 (Crandall-Rabinowitz [18]) *Let u^* be the extremal solution of (1.55) with $f(u) = e^u$ or $f(u) = (1+u)^p$, $p > 1$. If $n \leq 9$, then $u^* \in L^\infty(\Omega)$.*

Outline of the proof (in the case $f(u) = e^u$) Use the equation in (1.55) for the classical solutions $u = u_\lambda$ ($\lambda < \lambda^*$), together with the stability condition (1.54) for the test function $\xi = e^{\alpha u} - 1$ (for a positive exponent α to be chosen later). More precisely, start from (1.54)—with f' replaced by $\lambda f'$—and to proceed with $\int_\Omega \alpha^2 e^{2\alpha u} |\nabla u|^2$, write $\alpha^2 e^{2\alpha u} |\nabla u|^2 = (\alpha/2)\nabla \left(e^{2\alpha u} - 1\right) \nabla u$, and integrate by parts to use (1.55). For every $\alpha < 2$, verify that this leads, after letting $\lambda \uparrow \lambda^*$, to $e^{u^*} \in L^{2\alpha+1}(\Omega)$. As a consequence, by Calderón-Zygmund theory and Sobolev embeddings, $u^* \in W^{2,2\alpha+1}(\Omega) \subset L^\infty(\Omega)$ if $2(2\alpha + 1) > n$. This requires that $n \leq 9$.   □

Notice that the nonlinearities $f(u) = e^u$ or $f(u) = (1+u)^p$ with $p > 1$ satisfy (1.56).

In the radial case $\Omega = B_1$ we have the following result.

Theorem 1.53 (Cabré-Capella [12]) *Let u^* be the extremal solution of (1.55). Assume that f satisfies (1.56) and that $\Omega = B_1$. If $1 \leq n \leq 9$, then $u^* \in L^\infty(B_1)$.*

As mentioned before, this theorem also holds for every $H_0^1(B_1)$ stable weak solution of (1.52), for any $f \in C^1$. Thus, in view of Proposition 1.51, the dimension $n \leq 9$ is optimal in this result.

We turn now to the nonradial case and we present the currently known results. First, in 2000 Nedev solved the case $n \leq 3$.

Theorem 1.54 (Nedev [32]) *Let f be convex and satisfy (1.56), and $\Omega \subset \mathbb{R}^n$ be a smooth bounded domain. If $n \leq 3$, then $u^* \in L^\infty(\Omega)$.*

In 2010, Nedev's result was improved to dimension four:

Theorem 1.55 (Cabré [8]; Villegas [37]) *Let f satisfy (1.56), $\Omega \subset \mathbb{R}^n$ be a smooth bounded domain, and $1 \leq n \leq 4$. If $n \in \{3, 4\}$ assume either that f is a convex nonlinearity or that Ω is a convex domain. Then, $u^* \in L^\infty(\Omega)$.*

For $3 \leq n \leq 4$, [8] requires Ω to be convex, while f needs not be convex. Some years later, S. Villegas [37] succeeded to use both [8] and [32] when $n = 4$ to remove the requirement that Ω is convex by further assuming that f is convex.

Open problem 1.56 For every Ω and for every f satisfying (1.56), is the extremal solution u^*—or, in general, H_0^1 stable weak solutions of (1.52)—always bounded in dimensions 5, 6, 7, 8, 9?

We recall that the answer to this question is affirmative when $\Omega = B_1$, by Theorem 1.53. We next sketch the proof of this radial result, as well as the regularity theorem in the nonradial case up to $n \leq 4$. In the case $n = 4$, we will need the following remarkable result.

Theorem 1.57 (Allard; Michael and Simon) *Let $M \subset \mathbb{R}^{m+1}$ be an immersed smooth m-dimensional compact hypersurface without boundary.*
Then, for every $p \in [1, m)$, there exists a constant $C = C(m, p)$ depending only on the dimension m and the exponent p such that, for every C^∞ function $v : M \to \mathbb{R}$,

$$\left(\int_M |v|^{p^*} \, dV \right)^{1/p^*} \leq C(m, p) \left(\int_M (|\nabla v|^p + |\mathscr{H} v|^p) \, dV \right)^{1/p}, \qquad (1.57)$$

where $\mathscr{H}$ is the mean curvature of M and $p^* = mp/(m - p)$.

This theorem dates from 1972 and has its origin in an important result of Miranda from 1967. It stated that (1.57) holds with $\mathscr{H} = 0$ if M is a minimal surface in $\mathbb{R}^{m+1}$. See the book [25] for a proof of Theorem 1.57.

Remark 1.58 Note that this Sobolev inequality contains a term involving the mean curvature of M on its right-hand side. This fact makes, in a remarkable way, that the constant $C(m, p)$ in the inequality does not depend on the geometry of the manifold M.

Outline of the proof (of Theorems 1.53, 1.54, and 1.55) For Theorem 1.54 the test function to be used is $\xi = h(u)$, for some h depending on f (as in the proof of Theorem 1.52).

Instead, for Theorems 1.53 and 1.55, the proofs start by writing the stability condition (1.54) for the test function $\xi = \tilde{c}\eta$, where $\eta|_{\partial\Omega} \equiv 0$. This was motivated by the analogous computation that we have presented for minimal surfaces right after Remark 1.18. Integrating by parts, one easily deduces that

$$\int_\Omega \left(\Delta\tilde{c} + f'(u)\tilde{c} \right) \tilde{c}\eta^2 \, dx \leq \int_\Omega \tilde{c}^2 |\nabla\eta|^2 \, dx. \qquad (1.58)$$

Next, a key point is to choose a function $\tilde{c}$ satisfying an appropriate equation for the linearized operator $\Delta + f'(u)$. In the radial case (Theorem 1.53) the choice of $\tilde{c}$ and the final choice of ξ are

$$\tilde{c} = u_r \quad \text{and} \quad \xi = u_r r (r^{-\alpha} - (1/2)^{-\alpha})_+,$$

where $r = |x|$, $\alpha > 0$, and ξ is later truncated near the origin to make it Lipschitz. The proof in the radial case is quite simple after computing the equation satisfied by u_r.

For the estimate up to dimension 4 in the nonradial case (Theorem 1.55), [8] takes

$$\tilde{c} = |\nabla u| \quad \text{and} \quad \xi = |\nabla u|\, \varphi(u), \tag{1.59}$$

where, in dimension $n = 4$, φ is chosen depending on the solution u itself.

We make the choice (1.59) and, in particular, we take $\tilde{c} = |\nabla u|$ in (1.58). It is easy to check that, in the set $\{|\nabla u| > 0\}$, we have

$$\left(\Delta + f'(u)\right)|\nabla u| = \frac{1}{|\nabla u|} \left(\sum_{i,j} u_{ij}^2 - \sum_i \left(\sum_j u_{ij} \frac{u_j}{|\nabla u|} \right)^2 \right). \tag{1.60}$$

Taking an orthonormal basis in which the last vector is the normal $\nabla u/|\nabla u|$ to the level set of u (through a given point $x \in \Omega$), and the other vectors are the principal directions of the level set at x, one easily sees that (1.60) can be written as

$$\left(\Delta + f'(u)\right)|\nabla u| = \frac{1}{|\nabla u|} \left(|\nabla_T |\nabla u||^2 + |A|^2 |\nabla u|^2 \right) \quad \text{in } \Omega \cap \{|\nabla u| > 0\}, \tag{1.61}$$

where $|A|^2 = |A(x)|^2$ is the squared norm of the second fundamental form of the level set of u passing through a given point $x \in \Omega \cap \{|\nabla u| > 0\}$, i.e., the sum of the squares of the principal curvatures of the level set. In the notation of the first section on minimal surfaces, $|A|^2 = c^2$. On the other hand, as in that section $\nabla_T = \delta$ denotes the tangential gradient to the level set. Thus, (1.61) involves geometric information of the level sets of u.

Therefore, using the stability condition (1.58), we conclude that

$$\int_{\{|\nabla u|>0\}} \left(|\nabla_T|\nabla u||^2 + |A|^2|\nabla u|^2 \right) \eta^2 \, dx \le \int_\Omega |\nabla u|^2 |\nabla \eta|^2 \, dx. \tag{1.62}$$

Let us define

$$T := \max_{\overline{\Omega}} u = \|u\|_{L^\infty(\Omega)} \quad \text{and} \quad \Gamma_s := \{x \in \Omega : u(x) = s\}$$

for $s \in (0, T)$.

We now use (1.62) with $\eta = \varphi(u)$, where φ is a Lipschitz function in $[0, T]$ with $\varphi(0) = 0$. The right hand side of (1.62) becomes

$$\int_\Omega |\nabla u|^2 |\nabla \eta|^2 \, dx = \int_\Omega |\nabla u|^4 \varphi'(u)^2 dx$$

$$= \int_0^T \left(\int_{\Gamma_s} |\nabla u|^3 \, dV_s \right) \varphi'(s)^2 \, ds,$$

by the *coarea formula*. Thus, (1.62) can be written as

$$\int_0^T \left(\int_{\Gamma_s} |\nabla u|^3 \, dV_s \right) \varphi'(s)^2 \, ds$$

$$\geq \int_{\{|\nabla u| > 0\}} \left(|\nabla_T |\nabla u||^2 + |A|^2 |\nabla u|^2 \right) \varphi(u)^2 dx$$

$$= \int_0^T \left(\int_{\Gamma_s \cap \{|\nabla u| > 0\}} \frac{1}{|\nabla u|} \left(|\nabla_T |\nabla u||^2 + |A|^2 |\nabla u|^2 \right) dV_s \right) \varphi(s)^2 \, ds$$

$$= \int_0^T \left(\int_{\Gamma_s \cap \{|\nabla u| > 0\}} \left(4 \left| \nabla_T |\nabla u|^{1/2} \right|^2 + \left(|A| |\nabla u|^{1/2} \right)^2 \right) dV_s \right) \varphi(s)^2 \, ds.$$

We conclude that

$$\int_0^T h_1(s)\varphi(s)^2 \, ds \leq \int_0^T h_2(s)\varphi'(s)^2 \, ds, \tag{1.63}$$

for all Lipschitz functions $\varphi : [0, T] \to \mathbb{R}$ with $\varphi(0) = 0$, where

$$h_1(s) := \int_{\Gamma_s} \left(4|\nabla_T |\nabla u|^{1/2}|^2 + \left(|A||\nabla u|^{1/2} \right)^2 \right) dV_s, \quad h_2(s) := \int_{\Gamma_s} |\nabla u|^3 \, dV_s$$

for every regular value s of u. We recall that, by Sard's theorem, almost every $s \in (0, T)$ is a regular value of u.

Inequality (1.63), with h_1 and h_2 as defined above, leads to a bound for T (that is, to an L^∞ estimate and hence to Theorem 1.55) after choosing an appropriate test function φ in (1.63). In dimensions 2 and 3 we can choose a simple function φ in (1.63) and use well known geometric inequalities about the curvature of manifolds (note that h_1 involves the curvature of the level sets of u). Instead, in dimension 4 we need to use the geometric Sobolev inequality of Theorem 1.57 on each level set of u. Note that $\mathcal{H}^2 \leq (n-1)|A|^2$. This gives the following lower bound for $h_1(s)$:

$$c(n) \left(\int_{\Gamma_s} |\nabla u|^{\frac{n-1}{n-3}} \right)^{\frac{n-3}{n-1}} \leq h_1(s).$$

Comparing this with $h_2(s)$, which appears in the right hand side of (1.63), we only know how to derive an L^∞-estimate for u (i.e., a bound on $T = \max u$) when the exponent $(n-1)/(n-3)$ in the above inequality is larger than or equal to the exponent 3 in $h_2(s)$. This requires $n \leq 4$. See [8] for details on how the proof is finished.

□

Acknowledgements The authors wish to thank Lorenzo Cavallina for producing the figures of this work.

The first author is member of the Barcelona Graduate School of Mathematics and is supported by MINECO grants MTM2014-52402-C3-1-P and MTM2017-84214-C2-1-P. He is also part of the Catalan research group 2017 SGR 1392.

The second author was partially supported by PhD funds of the Università di Firenze and he is a member of the Gruppo Nazionale Analisi Matematica Probabilitá e Applicazioni (GNAMPA) of the Istituto Nazionale di Alta Matematica (INdAM). This work was partially written while the second author was visiting the Departament de Matemàtiques of the Universitat Politècnica de Catalunya, that he wishes to thank for hospitality and support.

Appendix: A Calibration Giving the Optimal Isoperimetric Inequality

Our first proof of Theorem 1.8 used a calibration. To understand better the concept and use of "calibrations", we present here another one. It leads to a proof of the isoperimetric problem.

The isoperimetric problems asks which sets in $\mathbb{R}^n$ minimize perimeter for a given volume. Making the first variation of perimeter (as in Sect. 1.1), but now with a volume constraint, one discovers that a minimizer Ω should satisfy $\mathcal{H} = c$ (with c a constant), at least in a weak sense, where $\mathcal{H}$ is the mean curvature of $\partial\Omega$. Obviously, balls satisfy this equation – they have constant mean curvature. The isoperimetric inequality states that the unique minimizers are, indeed, balls. In other words, we have:

Theorem 1.59 (The isoperimetric Inequality) *We have*

$$\frac{|\partial\Omega|}{|\Omega|^{\frac{n-1}{n}}} \geq \frac{|\partial B_1|}{|B_1|^{\frac{n-1}{n}}} \tag{1.64}$$

for every bounded smooth domain $\Omega \subset \mathbb{R}^n$. In addition, if equality holds in (1.64), then Ω must be a ball.

In 1996 the first author found the following proof of the isoperimetric problem. It uses a calibration (for more details see [10]).

Outline of the proof (of the isoperimetric inequality) The initial idea was to charac-
terize the perimeter $|\partial\Omega|$ as in (1.9)–(1.10), that is, as

$$|\partial\Omega| = \sup_{\|X\|_{L^\infty} \leq 1} \int_{\partial\Omega} X \cdot v \, dH_{n-1}.$$

Taking X to be a gradient, we have that

$$|\partial\Omega| = \int_{\partial\Omega} \nabla u \cdot v \, dH_{n-1} = \int_{\partial\Omega} u_v \, dH_{n-1},$$

for every function u such that $u_v = 1$ on $\partial\Omega$. Let us take u to be the solution of

$$\begin{cases} \Delta u = c & \text{in } \Omega \\ u_v = 1 & \text{on } \partial\Omega, \end{cases} \tag{1.65}$$

where c is a constant that, by the divergence theorem, is given by

$$c = \frac{|\partial\Omega|}{|\Omega|}.$$

It is known that there exists a unique solution u to (1.65) (up to an additive constant).
    Now let us see that $X = \nabla u$ (where X was the notation that we used in the
proof of Theorem 1.8) can play the role of a calibration. In fact, in analogy with
Definition 1.10 we have:

(i-bis) $\operatorname{div}\nabla u = \frac{|\partial\Omega|}{|\Omega|}$ in Ω;
(ii-bis) $\nabla u \cdot v = 1$ on $\partial\Omega$;
(iii-bis) $B_1(0) \subset \nabla u(\Gamma_u)$, where

$$\Gamma_u = \left\{ x \in \Omega : u(y) \geq u(x) + \nabla u(x) \cdot (y - x) \text{ for every } y \in \overline{\Omega} \right\}$$

is the *lower contact set of u*, that is, the set of the points of Ω at which the
tangent plane to u stays below u in Ω.

The relations (i-bis) and (ii-bis) follow immediately from (1.65). In the following
exercise, we ask to establish (iii-bis) and finish the proof of (1.64).
    We point out that this proof also gives that Ω must be a ball if equality holds
in (1.64).   □

Exercise 1.60 Establish (iii-bis) above. For this, use a foliation-contact argument
(as in the alternative proof of Theorem 1.8 and in the proof of Theorem 1.32),
foliating now $\mathbb{R}^n \times \mathbb{R}$ by parallel hyperplanes.
    Next, finish the proof of (1.64). For this, consider the measures of the two sets in
(iii-bis), compute $|\nabla u(\Gamma_u)|$ using the *area formula*, and control $\det D^2 u$ using the
geometric-arithmetic means inequality.

References

1. G. Alberti, L. Ambrosio, X. Cabré, On a long-standing conjecture of E. De Giorgi: symmetry in 3D for general nonlinearities and a local minimality property. Acta Appl. Math. **65**, 9–33 (2001)
2. L. Ambrosio, X. Cabré, Entire solutions of semilinear elliptic equations in $\mathbb{R}^3$ and a conjecture of De Giorgi. J. Am. Math. Soc. **13**, 725–739 (2000)
3. H. Berestycki, L. Caffarelli, L. Nirenberg, Further qualitative properties for elliptic equations in unbounded domains. Ann. Scuola Norm. Sup. Pisa Cl. Sci. **25**, 69–94 (1997)
4. H. Berestycki, L. Nirenberg, S.R.S. Varadhan, The principal eigenvalue and maximum principle for second-order elliptic operators in general domains. Commun. Pure Appl. Math. **47**, 47–92 (1994)
5. E. Bombieri, E. De Giorgi, E. Giusti, Minimal cones and the Bernstein problem. Inv. Math. **7**, 243–268 (1969)
6. H. Brezis, Is there failure of the inverse function theorem?, in *Morse Theory, Minimax Theory and Their Applications to Nonlinear Differential Equations*, vol. 1 (International Press, Somerville, 2003), pp. 23–33
7. X. Cabré, Extremal solutions and instantaneous complete blow-up for elliptic and parabolic problems, in *Perspectives in Nonlinear Partial Differential Equations: In honor of Haim Brezis*. Contemporary Mathematics (American Mathematical Society, Providence, 2007)
8. X. Cabré, Regularity of minimizers of semilinear elliptic problems up to dimension 4. Commun. Pure Appl. Math. **63**, 1362–1380 (2010)
9. X. Cabré, Uniqueness and stability of saddle-shaped solutions to the Allen-Cahn equation. J. Math. Pures Appl. **98**, 239–256 (2012)
10. X. Cabré, Isoperimetric, Sobolev, and eigenvalue inequalities via the Alexandroff-Bakelman-Pucci method: a survey. Chin. Ann. Math. Ser. B **38**, 201–214 (2017)
11. X. Cabré, Boundedness of stable solutions to semilinear elliptic equations: a survey. Adv. Nonlinear Stud. **17**, 355–368 (2017)
12. X. Cabré, A. Capella, Regularity of radial minimizers and extremal solutions of semilinear elliptic equations. J. Funct. Anal. **238**, 709–733 (2006)
13. X. Cabré, A. Capella, Regularity of minimizers for three elliptic problems: minimal cones, harmonic maps, and semilinear equations. Pure Appl. Math. Q. **3**, 801–825 (2007)
14. L.A. Caffarelli, D. Jerison, C.E. Kenig, Global energy minimizers for free boundary problems and full regularuty in three dimensions. Contemp. Math. **350**, 83–97 (2004)
15. L.A. Caffarelli, J.-M. Roquejoffre, O. Savin, Nonlocal minimal surfaces. Commun. Pure Appl. Math. **63**, 1111–1144 (2010)
16. T.H. Colding, W.P. Minicozzi II, *A Course in Minimal Surfaces*. Graduate Studies in Mathematics, vol. 121 (American Mathematical Society, Providence, 2011)
17. M. Cozzi, A. Figalli, Regularity theory for local and nonlocal minimal surfaces: an overview, in *Nonlocal and Nonlinear Diffusions Interaction: New Methods and Directions*. Lecture Notes in Mathematics. C.I.M.E. Foundation Subseries, vol. 2186 (Springer, Cham, 2017), pp. 117–158
18. M.G. Crandall, P.H. Rabinowitz, Some continuation and variational methods for positive solutions of nonlinear elliptic eigenvalue problems. Arch. Rat. Mech. Anal. **58**, 207–218 (1975)
19. A. Davini, On calibrations for Lawson's cones. Rend. Semin. Mat. Univ. Padova **111**, 55–70 (2004)
20. E. De Giorgi, Convergence problems for functionals and operators, in *Proc. Int. Meeting on Recent Methods in Nonlinear Analysis (Rome, 1978)*, Pitagora, Bologna (1979), pp. 131–188
21. M. del Pino, M. Kowalczyk, J. Wei, On De Giorgi's conjecture in dimension $N \geq 9$. Ann. Math. **174**, 1485–1569 (2011)
22. D. De Silva, D. Jerison, A singular energy minimizing free boundary. J. Reine Angew. Math. **635**, 1–22 (2009)
23. G. De Philippis, E. Paolini, A short proof of the minimality of Simons cone. Rend. Semin. Mat. Univ. Padova **121**, 233–242 (2009)

24. S. Dipierro, E. Valdinoci, Nonlocal minimal surfaces: interior regularity, quantitative estimates and boundary stickiness (2016). Preprint. arxiv:1607.06872
25. L. Dupaigne, *Stable Solutions of Elliptic Partial Differential Equations*. Monographs and Surveys in Pure and Applied Mathematics, vol. 143 (Chapman & Hall/CRC, Boca Raton, FL, 2011)
26. N. Ghoussoub, C. Gui, On a conjecture of De Giorgi and some related problems. Math. Ann. **311**, 481–491 (1998)
27. M. Giaquinta, J. Souček, Harmonic maps into a hemisphere. Ann. Scuola Norm. Sup. Pisa Cl. Sci. (4) **12**, 81–90 (1985)
28. E. Giusti, *Minimal Surfaces and Functions of Bounded Variation*. Monographs in Mathematics, vol. 80 (Birkhäuser Verlag, Basel, 1984)
29. W. Jäger, H. Kaul, Rotationally symmetric harmonic maps from a ball into a sphere and the regularity problem for weak solutions of elliptic systems. J. Reine Angew. Math. **343**, 146–161 (1983)
30. D. Jerison, O. Savin, Some remarks on stability of cones for the one-phase free boundary problem. Geom. Funct. Anal. **25**, 1240–1257 (2015)
31. Y. Liu, K. Wang, J. Wei, Global minimizers of the Allen Cahn equation in dimension $n \geq 8$. J. Math. Pures Appl. **108**, 818–840 (2017)
32. G. Nedev, Regularity of the extremal solution of semilinear elliptic equations. C. R. Acad. Sci. Paris **330**, 997–1002 (2000)
33. O. Savin, Regularity of flat level sets in phase transitions. Ann. Math. **169**, 41–78 (2009)
34. O. Savin, E. Valdinoci, Regularity of nonlocal minimal cones in dimension 2. Calc. Var. Partial Differ. Equ. **48**, 33–39 (2013)
35. R. Schoen, K. Uhlenbeck, Regularity of minimizing harmonic maps into the sphere. Invent. Math. **78**, 89–100 (1984)
36. J. Simons, Minimal varieties in riemannian manifolds. Ann. Math. **88**, 62–105 (1968)
37. S. Villegas, Boundedness of extremal solutions in dimension 4. Adv. Math. **235**, 126–133 (2013)

Chapter 2
Isoperimetric Inequalities for Eigenvalues of the Laplacian

Antoine Henrot

Abstract These lecture notes give an overview of "isoperimetric inequalities", namely inequalities involving only geometric features, for the eigenvalues of the Laplace operator, with Dirichlet boundary conditions. In other words, we are mainly interested in minimization problems like

$$\min\{\lambda_k(\Omega), \ \Omega \subset \mathbb{R}^N \text{ open set, with some geometric constraints. }\} \qquad (2.1)$$

Here $\lambda_k(\Omega)$ denotes the k-th eigenvalue of the Laplace operator with Dirichlet boundary conditions and the geometric constraints can involve the volume or the perimeter or the diameter or some box constraints or some specific sub-classes like polygons or convex sets. Most of the information contained in these notes are from the book of the author (Henrot, Extremum problems for eigenvalues of elliptic operators. Frontiers in mathematics, Birkhäuser, Basel, 2006), but some more recent results are also presented.

These lecture notes give an overview of "isoperimetric inequalities", namely inequalities involving only geometric features, for the eigenvalues of the Laplace operator, with Dirichlet boundary conditions. In other words, we will be mainly interested in minimization problems like

$$\min\{\lambda_k(\Omega), \ \Omega \subset \mathbb{R}^N \text{ open set, with some geometric constraints. }\} \qquad (2.2)$$

Here $\lambda_k(\Omega)$ denotes the k-th eigenvalue of the Laplace operator (see Notation in Sect. 2.1) and the geometric constraints can involve the volume or the perimeter or some box constraints or some specific sub-classes like polygons or convex sets. Most of the information contained in these notes are from the book of the author

A. Henrot (✉)
Institut Élie Cartan, UMR CNRS 7502, Université de Lorraine, Vandœuvre-lès-Nancy, France
e-mail: antoine.henrot@univ-lorraine.fr

© Springer Nature Switzerland AG 2018
C. Bianchini et al. (eds.), *Geometry of PDEs and Related Problems*,
Lecture Notes in Mathematics 2220, https://doi.org/10.1007/978-3-319-95186-7_2

[29], but some more recent results are also presented for which a good recent reference is [30]. Another classical reference is [15].

These notes are organized as follows: Sect. 2.1 recall the basic facts on the eigenvalues for an elliptic operator. We are only concerned here by the Laplacian as a model operator. This section also fix notation. In Sect. 2.2, we consider various problems involving the first eigenvalue of the Laplace operator with Dirichlet boundary conditions. We start with the very classical Rayleigh-Faber-Krahn inequality and we present a (recent) quantitative version of it. We also deal with the same minimization problem in the subclass of polygons, presenting the famous Pòlya conjecture. We report on some obstacle problems (finding the optimal obstacle with various constraints) and we also give some results on the minimization problem with a box constraint. Then, in Sect. 2.3, we consider the second eigenvalue. We prove the Hong-Krahn-Szego Theorem (with a quantitative version) and we also consider the problem of minimizing λ_2 among convex domains. At last, in Sect. 2.4, we report on the minimization problems for $\lambda_k(\Omega), k \geq 3$ with a volume, a perimeter or a diameter constraint. All the topics we discuss are the occasion to present classical tools of modern analysis and shape optimization like

- Rearrangement (Schwarz and Steiner symmetrization)
- Hausdorff convergence (of compact and open sets)
- γ-convergence and convergence of eigenvalues
- Domain derivative.

We also give a collection of open problems (easy to state, difficult to solve) showing that this field is still very active.

2.1 Notation and Prerequisites

In this section, we recall the basic properties of the eigenvalues of the Laplace operator. All the results that we state here are also valid for more general (linear) elliptic operators. For the proofs or more details, we refer to any textbook on partial differential equations and operator theory, as [10, 21, 24] which are good standard references.

2.1.1 Notation and Sobolev Spaces

Let Ω be a bounded open set in $\mathbb{R}^N$, its boundary is denoted by $\partial\Omega$. We denote by $|\Omega|$ the Lebesgue measure (or volume) of Ω, by $P(\Omega)$ its perimeter (for example in the sense of De Giorgi) and by $D(\Omega)$ its diameter. We denote by $L^2(\Omega)$ the Hilbert space of square integrable functions defined on Ω and by $H^1(\Omega)$ the Sobolev space of functions in $L^2(\Omega)$ whose partial derivatives (in the sense of distributions) are in

$L^2(\Omega)$:

$$H^1(\Omega) := \{u \in L^2(\Omega) \text{ such that } \frac{\partial u}{\partial x_i} \in L^2(\Omega), \ i = 1, 2, \ldots, N\}.$$

This is an Hilbert space when it is endowed with the scalar product

$$(u, v)_{H^1} := \int_\Omega u(x)v(x)\, dx + \int_\Omega \nabla u(x).\nabla v(x)\, dx$$

and the corresponding norm:

$$\|u\|_{H^1} := \left(\int_\Omega u(x)^2\, dx + \int_\Omega |\nabla u(x)|^2\, dx\right)^{1/2}.$$

Since we are always dealing with Dirichlet boundary conditions in these notes, we introduce the subspace $H_0^1(\Omega)$ which is defined as the closure of C^∞ functions compactly supported in Ω (functions in $C_0^\infty(\Omega)$) for the norm $\|\ \|_{H^1}$. It is also a Hilbert space.

By definition, $H_0^1(\Omega)$ and $H^1(\Omega)$ are continuously embedded in $L^2(\Omega)$, but we will need later a compact embedding. This is the purpose of the following theorem.

Theorem 2.1 (Rellich) *For any bounded open set Ω, the embedding $H_0^1(\Omega) \hookrightarrow L^2(\Omega)$ is compact.*

2.1.2 Eigenvalues and Eigenfunctions

2.1.2.1 Abstract Spectral Theory

Let us now give the abstract theorem which provides the existence of a sequence of eigenvalues and eigenfunctions. Let H be a Hilbert space endowed with a scalar product $(.,.)$, we recall that an operator T is a linear continuous map from H into H. We say that:

- T is positive if, $\forall x \in H$, $(Tx, x) \geq 0$,
- T is self-adjoint, if $\forall x, y \in H$, $(Tx, y) = (x, Ty)$,
- T is compact, if the image of any bounded set is relatively compact (i.e. has a compact closure) in H.

Theorem 2.2 *Let H be a separable Hilbert space of infinite dimension and T a self-adjoint, compact and positive operator. Then, there exists a sequence of real positive eigenvalues (v_n), $n \geq 1$ converging to 0 and a sequence of eigenvectors (x_n), $n \geq 1$ defining an Hilbert basis of H such that $\forall n$, $T x_n = v_n x_n$.*

Of course, this theorem can be seen as a generalization to Hilbert spaces of the classical result in finite dimension for symmetric or normal matrices (existence of real eigenvalues and of an orthonormal basis of eigenvectors).

2.1.2.2 Application to the Laplacian

We apply Theorem 2.2 to $H = L^2(\Omega)$ and the operator A (called the resolvent operator) defined by:

$$A_\Omega : L^2(\Omega) \to H_0^1(\Omega) \subset L^2(\Omega)$$
$$f \mapsto u \text{ solution of (2.4).} \tag{2.3}$$

where (2.4) is the variational formulation of the partial differential equation $-\Delta u = f$ in Ω with the boundary condition $u = 0$ on $\partial\Omega$:

$$\begin{cases} u \in H_0^1(\Omega) \text{ and } \forall v \in H_0^1(\Omega), \\ \int_\Omega \nabla u . \nabla v \, dx = \int_\Omega f v(x) \, dx . \end{cases} \tag{2.4}$$

Existence and uniqueness of a solution for problem (2.4) for a bounded open set Ω follows from Lax-Milgram Theorem and the Poincaré inequality. It is easy to check that A_Ω is positive, self-adjoint and compact (by using Rellich Theorem). Therefore Theorem 2.2 applies: there exists (u_n) a Hilbert basis of $L^2(\Omega)$ and a sequence $v_n > 0$, converging to 0, such that $A u_n = v_n u_n$.

Setting $\lambda_n = \frac{1}{v_n}$, we have proved:

Theorem 2.3 *Let Ω be a bounded open set in $\mathbb{R}^N$. There exists a sequence of positive eigenvalues (going to $+\infty$) and a sequence of corresponding eigenfunctions (defining an Hilbert basis of $L^2(\Omega)$) that we will denote respectively $0 < \lambda_1(\Omega) \le \lambda_2(\Omega) \le \lambda_3(\Omega) \le \dots$ and $u_1, u_2, u_3, \dots$ satisfying:*

$$\begin{cases} -\Delta u_n = \lambda_n(\Omega) u_n & \text{in } \Omega \\ u_n = 0 & \text{on } \partial\Omega . \end{cases} \tag{2.5}$$

Since the eigenfunctions are defined up to a multiplicative constant, we generally decide to normalize the eigenfunctions by the condition

$$\int_\Omega u_n(x)^2 \, dx = 1 . \tag{2.6}$$

Of course, it can occur that some eigenvalues are multiple (especially when the domain has symmetries). In this case, the eigenvalues are counted with their multiplicity.

Remark 2.4 When Ω is non connected, for example if Ω has two connected components $\Omega = \Omega_1 \cup \Omega_2$, we obtain the eigenvalues of Ω by collecting and reordering the eigenvalues of each connected components:

$$\lambda_1(\Omega) = \min(\lambda_1(\Omega_1), \lambda_1(\Omega_2))$$
$$\lambda_2(\Omega) = \min(\max(\lambda_1(\Omega_1), \lambda_1(\Omega_2)), \lambda_2(\Omega_1), \lambda_2(\Omega_2)) \qquad (2.7)$$
$$\vdots$$

More generally, we can always choose every eigenfunction of a disconnected open set Ω to vanish on all but one of the connected components of Ω. In particular, when the two connected components are the same, we will have $\lambda_1(\Omega) = \lambda_2(\Omega)$, i.e. λ_1 is a *double* eigenvalue.

That cannot happen when Ω is connected:

Theorem 2.5 *Let us assume that Ω is a regular connected open set. Then the first eigenvalue $\lambda_1(\Omega)$ is simple and the first eigenfunction u_1 has a constant sign on Ω. Usually, we choose it to be positive on Ω.*

2.1.3   Properties of Eigenvalues

Since the Laplacian is invariant for translations and rotations, for any isometry R, we have

$$\lambda_n(R(\Omega)) = \lambda_n(\Omega) . \qquad (2.8)$$

In the same way, it is immediate to check that, if H_k denotes a dilation (or homothety) of ratio $k > 0$:

$$\lambda_n(H_k(\Omega)) = \frac{\lambda_n(\Omega)}{k^2} . \qquad (2.9)$$

An important consequence of (2.9) is the following. In the sequel, we will often consider minimization problems with a volume constraint, like

$$\min\{\lambda_n(\Omega), \ |\Omega| = c\} . \qquad (2.10)$$

Then, it is often convenient to replace Problem (2.10) by:

$$\min \ |\Omega|^{2/N} \lambda_n(\Omega), \ . \qquad (2.11)$$

since these two problems (2.10) and (2.11) are clearly equivalent.

2.1.4 Some Examples

In this section, we are interested in the eigenvalues of the Laplacian for some very simple domains.

2.1.4.1 Rectangles

For rectangles, using the classical trick of separation of variables, we prove

Proposition 2.6 *Let $\Omega = (0, L) \times (0, l)$ be a plane rectangle, then its eigenvalues and eigenfunctions for the Laplacian with Dirichlet boundary conditions are:*

$$\begin{aligned} \lambda_{m,n} &= \pi^2 \left(\frac{m^2}{L^2} + \frac{n^2}{l^2} \right) \\ u_{m,n}(x, y) &= \frac{2}{\sqrt{Ll}} \sin(\frac{m\pi x}{L}) \sin(\frac{n\pi y}{l}) \end{aligned} \quad m, n \geq 1, \tag{2.12}$$

It is immediate to check that the pair $(\lambda_{m,n}, u_{m,n})$ given by (2.12) defines an eigenvalue and an eigenfunction for the Laplacian with Dirichlet boundary condition. Of course, the difficulty is to prove that there are no other possibility. Actually, it is due to the fact that the functions $\sin(\frac{m\pi x}{L}) \sin(\frac{n\pi y}{l})$ $m, n \geq 1$ form a complete orthogonal system in $L^2(\Omega)$, see [21].

Exercise 2.7 Solve the following minimization problems for the three different constraints and for $k = 1, 2, 3$:

$$\min\{\lambda_k(\Omega), \Omega \text{ rectangle}, |\Omega| = 1\}, \tag{2.13}$$

$$\min\{\lambda_k(\Omega), \Omega \text{ rectangle}, P(\Omega) = 4\}, \tag{2.14}$$

$$\min\{\lambda_k(\Omega), \Omega \text{ rectangle}, D(\Omega) = \sqrt{2}\}. \tag{2.15}$$

Is the solution always unique?
Can you state conjectures?

For the behaviour of the optimal rectangle (with an area constraint) when $k \to +\infty$ and relation with the famous lattice problem, see [4]. In this paper, the authors prove that, as k goes to infinity, the optimal rectangle approaches the square, see also Theorem 2.68 below.

2.1.4.2 Disks

Let us consider the disk B_R of radius R centered at O. Working in polar coordinates (r, θ), leads us to solve an ordinary differential equation in r known as the Bessel equation. Then, we can state

Proposition 2.8 *Let* $\Omega = B_R$ *be a disk of radius* R, *then its eigenvalues and eigenfunctions for the Laplacian with Dirichlet boundary conditions are:*

$$
\begin{aligned}
&\lambda_{0,k} = \frac{j_{0,k}^2}{R^2},\ k \geq 1, \\
&u_{0,k}(r,\theta) = \sqrt{\frac{1}{\pi}} \frac{1}{R|J_0'(j_{0,k})|}\, J_0(j_{0,k}r/R),\ k \geq 1, \\
&\lambda_{n,k} = \frac{j_{n,k}^2}{R^2},\ n,k \geq 1,\ \text{double eigenvalue} \\
&u_{n,k}(r,\theta) = \begin{matrix} \sqrt{\frac{2}{\pi}} \frac{1}{R|J_n'(j_{n,k})|}\, J_n(j_{n,k}r/R)\cos n\theta \\ \sqrt{\frac{2}{\pi}} \frac{1}{R|J_n'(j_{n,k})|}\, J_n(j_{n,k}r/R)\sin n\theta \end{matrix},\ n,k \geq 1,
\end{aligned}
\tag{2.16}
$$

where $j_{n,k}$ *is the k-th zero of the Bessel function* J_n.

Remark 2.9 Similarly, in dimension $N \geq 3$, the eigenvalues of the ball B_R of radius R involve the zeros of the Bessel functions $J_{N/2-1}$, $J_{N/2}$, …. For example

$$
\lambda_1(B_R) = \frac{j_{N/2-1,1}^2}{R^2} \qquad \lambda_2(B_R) = \lambda_3(B_R) = \ldots = \lambda_{N+1}(B_R) = \frac{j_{N/2,1}^2}{R^2}
\tag{2.17}
$$

2.1.5 Min-Max Principles and Applications

One very useful tool is the following variational characterization of the eigenvalues, known as Poincaré principle or Courant-Fischer formulae, see [21]:

$$
\lambda_k(\Omega) = \min_{\substack{E_k \subset H_0^1(\Omega), \\ \text{subspace of dim } k}} \max_{v \in E_k, v \neq 0} \frac{\int_\Omega |\nabla v|2\, dx}{\int_\Omega v^2\, dx},
\tag{2.18}
$$

The quantity $\frac{\int_\Omega |\nabla v|2\, dx}{\int_\Omega v^2\, dx}$ is called the **Rayleigh quotient** of the function v. In formulae (2.18), the minimum is achieved for choosing E_k the space spanned by the k-th first eigenfunctions. In particular, for the first eigenvalue, we get

$$
\lambda_1(\Omega) = \min_{v \in H_0^1(\Omega), v \neq 0} \frac{\int_\Omega |\nabla v(x)|^2\, dx}{\int_\Omega v(x)^2\, dx}.
\tag{2.19}
$$

In (2.19), the minimum is achieved by the corresponding eigenfunction.

2.1.5.1 Monotonicity

Let us consider two open bounded sets such that $\Omega_1 \subset \Omega_2$. This inclusion induces a natural embedding $H_0^1(\Omega_1) \hookrightarrow H_0^1(\Omega_2)$ just by extending by zero functions in $H_0^1(\Omega_1)$. In particular, the min-max principle implies the following monotonicity for inclusion for the eigenvalues with Dirichlet boundary conditions:

$$\Omega_1 \subset \Omega_2 \implies \lambda_k(\Omega_1) \geq \lambda_k(\Omega_2) \tag{2.20}$$

(since the minimum is taken over a larger class for $\lambda_k(\Omega_2)$). Moreover, the inequality is strict as soon as $\Omega_2 \setminus \Omega_1$ contains a set of positive capacity (since the first eigenfunction cannot vanish on such a set).

2.1.5.2 Nodal Domains

Let us now have a look to the sign of eigenfunctions. We have already seen in Theorem 2.5 that the first eigenfunction u_1 is positive in Ω when Ω is connected. More generally, u_1 is non-negative (positive on one connected component and it vanishes on the other ones). Actually, we can recover this result thanks to the minimum formulae (2.18) by using the fact that if $u \in H_0^1(\Omega)$ we also have $|u| \in H_0^1(\Omega)$ and we see that u and $|u|$ have the same Rayleigh quotient. Therefore, $|u_1|$ is also a minimizer of the Rayleigh quotient and, therefore, an eigenfunction.

Concerning the other eigenfunctions, since there are all orthogonal to u_1, they have to change of sign in Ω.

Definition 2.10 Let u_k, $k \geq 2$ be an eigenfunction. The connected components of the open sets

$$\Omega_+ = \{x \in \Omega,\, u_k(x) > 0\} \quad \text{and} \quad \Omega_- = \{x \in \Omega,\, u_k(x) < 0\}$$

are called **the nodal domains** of u_k.

The number of these nodal domains is bounded from above:

Theorem 2.11 (Nodal Domains) *Let u_k, $k \geq 2$ be a k-th eigenfunction. Then, u_k has at most k nodal domains.*

The proof consists in assuming that u_k has more than k nodal domains, then constructing a test function, orthogonal to the $(k-1)$-th first eigenfunctions, whose Rayleigh quotient has a value strictly less than λ_k to reach a contradiction by applying the min-max formulae, see e.g. [21]. Let us remark that Theorem 2.11 is also true for $k = 1$ and that it gives an elementary proof of the non-negativity of the first eigenfunction (see Theorem 2.5) without regularity assumptions. Moreover, it also implies that the first eigenfunction must be simple in the connected case since two non negative and non zero functions cannot be orthogonal.

We will also frequently use the following property of a nodal domain:

Proposition 2.12 *Let u_k, $k \geq 2$ be an eigenfunction of the Laplacian with Dirichlet boundary conditions associated with the eigenvalue λ_k. Let ω_k be one of its nodal domain. Then*

$$\lambda_1(\omega_k) = \lambda_k .$$

Indeed, since u_k satisfies $-\Delta u_k = \lambda_k u_k$ in ω_k and vanishes on $\partial \omega_k$, it is an eigenfunction for Δ on ω_k with Dirichlet boundary condition. Moreover, since u_k has a constant sign on ω_k, it can only be the first one.

Let us be a little bit more precise for the second Dirichlet-eigenfunction u_2 of the Laplacian. According to Theorem 2.11, u_2 has at most two nodal domains. So, it has exactly two nodal domains when Ω is connected. The set

$$\mathcal{N} = \{x \in \Omega, \, u_2(x) = 0\}$$

is called **the nodal line** of u_2. When Ω is a plane convex domain, this nodal line hits the boundary of Ω at exactly two points, see Melas [46], or Alessandrini [1]. For general simply connected plane domains Ω, it is still a conjecture, named after Larry Payne, the "Payne conjecture".

2.1.6 Topological Derivative

When a domain minimizes a given function of eigenvalues, we can classically get some optimality conditions by letting the boundary of the domain vary and using domain derivative formulae which will be presented in the Tool D below. An alternative way to get optimality conditions *inside* the domain is to use asymptotic expansions of eigenvalues of domains with small holes (which is often called *topological derivative* see e.g. [47]). There is a huge literature on that topic, see e.g. [26, 44, 50] and the references therein. Let us give an example:

Theorem 2.13 *Let Ω be an open set in $\mathbb{R}^N$, $x_0 \in \Omega$ and $\varepsilon > 0$ a small number. Let us denote by $\Omega_\varepsilon = \Omega \setminus B(x_0, \varepsilon)$ (the set where we have removed the ball centered at x_0 of radius ε). Then the simple eigenvalues of the Laplacian-Dirichlet operator satisfy the following expansion:*

$$\begin{aligned} \lambda_k(\Omega_\varepsilon) &= \lambda_k(\Omega) + \tfrac{2\pi}{-\log \varepsilon} u_k^2(x_0) + o(\tfrac{1}{|\log \varepsilon|}) && \text{if } N = 2 \\ \lambda_k(\Omega_\varepsilon) &= \lambda_k(\Omega) + \varepsilon^{N-2}(N-2)S^{N-1} u_k^2(x_0) + o(\varepsilon^{N-2}) && \text{if } N \geq 3 \end{aligned} \quad (2.21)$$

where, in the last formulae, S^{N-1} is the $N-1$-dimensional measure of the unit sphere in $\mathbb{R}^N$.

The previous formulae can be possibly used to prove *non existence* of a minimizer for some function of eigenvalues.

2.2 The First Eigenvalue

2.2.1 The Faber-Krahn Inequality

For the first eigenvalue, the basic result is, as conjectured by Lord Rayleigh and later proved by G. Faber and E. Krahn, see [25, 39, 40]:

Theorem 2.14 (Faber-Krahn) *Let c be a positive number and B the ball of volume c. Then,*

$$\lambda_1(B) = \min\{\lambda_1(\Omega), \ \Omega \text{ open subset of } \mathbb{R}^N, \ |\Omega| = c\}.$$

The classical proof makes use of the Schwarz (spherical decreasing) rearrangement that we describe now.

Tool A: Schwarz Rearrangement

The reader who wants more details (including other types of rearrangement) can read for example [5, 27, 36, 38, 52].

Definition 2.15 For any measurable set ω in $\mathbb{R}^N$, we denote by ω^* **the ball of same volume as** ω.

If u is a non negative measurable function defined on a measurable set Ω and vanishing on its boundary $\partial\Omega$, we denote by $\Omega(c) = \{x \in \Omega \, /u(x) \geq c\}$ its level sets.

The Schwarz rearrangement (or spherical decreasing rearrangement) of u is the function u^* defined on Ω^* by

$$u^*(x) = \sup\{c \, /x \in \Omega(c)^*\} \, .$$

In other words, u^* is constructed from u by rearranging the level sets of u in balls of same volume. By construction, the following properties of u^* are obvious:

- u^* is radially symmetric, non increasing as a function of $|x|$,
- $\sup_\Omega u = \sup_\Omega u^*$,
- u and u^* are equimeasurable (i.e. their level sets have same measure).

As an immediate consequence of the last point and the Cavalieri principle, we have:

Theorem 2.16 *Let Ω be a measurable set and u be a non negative measurable function defined on Ω and vanishing on its boundary $\partial\Omega$. Let ψ be any measurable function defined on $\mathbb{R}^+$ with values in $\mathbb{R}$, then*

$$\int_\Omega \psi(u(x)) \, dx = \int_{\Omega^*} \psi(u^*(x)) \, dx \, . \tag{2.22}$$

Let us now state a deeper result, sometimes called the Pòlya inequality, which gives a connection between the integrals of the gradients of u and u^*. Its proof relies on the classical isoperimetric inequality, see the above references.

Theorem 2.17 (Pòlya's Inequality) *Let Ω be an open set and u a non negative function belonging to the Sobolev space $H_0^1(\Omega)$. Then $u^* \in H_0^1(\Omega^*)$ and*

$$\int_\Omega |\nabla u(x)|^2\, dx \geq \int_{\Omega^*} |\nabla u^*(x)|^2\, dx \ . \tag{2.23}$$

Another useful inequality is due to Hardy and Littlewood.

Theorem 2.18 (Hardy-Littlewood) *Let u and v be two non negative measurable functions defined on Ω and u^*, v^* their respective spherical decreasing rearrangements. Then*

$$\int_\Omega u(x)v(x)dx \leq \int_{\Omega^*} u^*(x)v^*(x)dx \ . \tag{2.24}$$

$\boxed{\text{end of Tool A}}$

Let us now apply this rearrangement technique to prove Theorem 2.2.1. Let Ω be a bounded open set of volume c and $\Omega^* = B$ the ball of same volume. Let u_1 denotes an eigenfunction associated to $\lambda_1(\Omega)$ and u_1^* its rearrangement. From (2.22) and (2.23), it comes:

$$\int_{\Omega^*} u_1^*(x)^2\, dx = \int_\Omega u_1(x)^2\, dx \quad \text{and} \quad \int_{\Omega^*} |\nabla u_1^*(x)|^2\, dx \leq \int_\Omega |\nabla u_1(x)|^2\, dx \tag{2.25}$$

Now, according to (2.19), we have

$$\lambda_1(\Omega^*) \leq \frac{\int_{\Omega^*} |\nabla u_1^*(x)|^2\, dx}{\int_{\Omega^*} u_1^*(x)^2\, dx} \quad \text{and} \quad \lambda_1(\Omega) = \frac{\int_\Omega |\nabla u_1(x)|^2\, dx}{\int_\Omega u_1(x)^2\, dx} \ . \tag{2.26}$$

Then, (2.25) together with (2.26) yields the desired result.

2.2.2 A Quantitative Version of Faber-Krahn Inequality

Can we improve the Faber-Krahn inequality, by proving its stability or giving a quantitative version of it? In other words, if we have a domain Ω of area π whose first eigenvalue is close to the first eigenvalue of the unit disk can we claim that Ω is close to this disk? In what sense? This kind of question has attracted much interest these last years, starting with the classical isoperimetric inequality. We refer to [30, Chapter 7] for a complete overview and many references. Let us describe here one of the main result. For that purpose, we define the *Fraenkel asymmetry* which measures the deviation with respect to a ball. It is nothing else that a L^1 distance between Ω and the closest ball of same volume:

$$\mathscr{A}(\Omega) = \inf\left\{ \frac{|\Omega \Delta B|}{|\Omega|} : B \text{ ball such that } |B| = |\Omega| \right\} \tag{2.27}$$

where $\Omega \Delta B$ denotes the symmetric difference between the two sets. Observe that for every ball B such that $|B| = |\Omega|$, we have $|\Omega \Delta B| = 2\,|\Omega \setminus B| = 2\,|B \setminus \Omega|$. Now we have, see [9] for the original paper

Theorem 2.19 (Brasco-de Philippis-Velichkov) *There exists a dimensional constant $\gamma_N > 0$ such that for every open $\Omega \subset \mathbb{R}^N$ with finite measure, we have*

$$|\Omega|^{\frac{2}{N}} \lambda_1(\Omega) - |B|^{\frac{2}{N}} \lambda_1(B) \geq \gamma_N \,\mathscr{A}(\Omega)^2. \tag{2.28}$$

Open Problem 2.20 *Find the value, or at least an estimate, of the optimal constant γ_N in (2.28).*

2.2.3 The Case of Polygons

We can now ask the question of minimizing λ_1 in the class of polygons with a given number of sides. We denote by $\mathscr{P}_N$ the class of plane polygons with *at most N* edges. We begin by an existence result.

Theorem 2.21 *Let $a > 0$ and $N \in \mathbb{N}$ be fixed. Then the problem*

$$\min\{\lambda_1(\Omega), \ \Omega \in \mathscr{P}_N, \ |\Omega| = a\} \tag{2.29}$$

has a solution. This one has exactly N edges.

The proof of existence follows the direct method of calculus of variations. For that purpose, we need to prove that the eigenvalues are continuous with respect to some basic convergence of domains. This is the aim of the next tools.

Tool B: Hausdorff Convergence

Let us recall the definition of the Hausdorff distance between two compact sets (Fig. 2.1).

Definition 2.22 Let K_1, K_2 be two non-empty compact sets in $\mathbb{R}^N$. We set

$$\forall x \in \mathbb{R}^N, \ d(x, K_1) := \inf_{y \in K_1} |y - x|$$
$$\rho(K_1, K_2) := \sup_{x \in K_1} d(x, K_2)$$

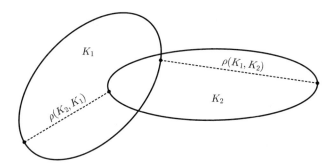

Fig. 2.1 Hausdorff distance of two compact sets: $d^H(K_1, K_2) := \max(\rho(K_1, K_2), \rho(K_2, K_1))$

Then the Hausdorff distance of K_1 and K_2 is defined by

$$d^H(K_1, K_2) := \max(\rho(K_1, K_2), \rho(K_2, K_1)). \tag{2.30}$$

We also have the two equivalent definitions:

$$d^H(K_1, K_2) = \inf\{\alpha > 0; \ K_2 \subset K_1^\alpha \ \text{and} \ K_1 \subset K_2^\alpha\} \tag{2.31}$$

where $K^\alpha = \{x \in \mathbb{R}^N; d(x, K) \le \alpha\}$,

$$d^H(K_1, K_2) = \|d_{K_1} - d_{K_2}\|_{L^\infty(\mathbb{R}^N)} = \|d_{K_1} - d_{K_2}\|_{L^\infty(K_1 \cup K_2)} \tag{2.32}$$

where, for any compact K, the function d_K is defined by $d_K(x) = d(x, K)$.

For open sets, we define the Hausdorff distance through their complementary:

Definition 2.23 Let Ω_1, Ω_2 be two open subsets of a (large) compact set B. Then their Hausdorff distance is defined by:

$$d_H(\Omega_1, \Omega_2) := d^H(B \setminus \Omega_1, B \setminus \Omega_2). \tag{2.33}$$

One of the most useful property of the Hausdorff distance is the following compactness property (see [15, 32]):

Theorem 2.24 *Let B be a fixed compact set in $\mathbb{R}^N$ and Ω_n a sequence of open subsets of B. Then, there exists an open set $\Omega \subset B$ and a subsequence Ω_{n_k} which converges for the Hausdorff distance to Ω.*

$$\boxed{\text{end of tool B}}$$

$$\boxed{\textbf{Tool C: } \gamma\textbf{-Convergence and Continuity of Eigenvalues}}$$

Let us begin with the definition of γ-convergence (for the Laplacian). While the Hausdorff convergence is purely geometric, this one has an analytic flavour.

Definition 2.25 Let D be a fixed ball, $\Omega_n \subset D$ a sequence of open sets and $\Omega \subset D$ an open set. We say that Ω_n γ-converge to Ω (and we write $\Omega_n \xrightarrow{\gamma} \Omega$) if, for every $f \in L^2(D)$, the solution u_n of $-\Delta u_n = f$ in Ω_n with $u_n = 0$ on $\partial \Omega_n$ converges (strongly) in $L^2(D)$ to u the solution of the same problem on Ω (as usual, every function in $H_0^1(\Omega_n)$ is extended by zero outside Ω_n).

In general, Hausdorff convergence defined above does not imply γ-convergence. Some more information is required. Let us now give the link with the convergence of eigenvalues. We refer to Sect. 2.1.2.2 for the definition of the resolvent operators A_Ω.

Theorem 2.26 *Let Ω_n and Ω open sets in a bounded box D. The following properties are equivalent:*

(i) Ω_n γ-converge to Ω.
(ii) (strong) convergence of resolvent operators: $\|A_{\Omega_n} - A_\Omega\| \to 0$.

Now, it is well known that strong convergence of operators implies convergence of eigenvalues. Therefore, we have

Corollary 2.27 *If Ω_n γ-converge to Ω, then for any k, $\lambda_k(\Omega_n) \to \lambda_k(\Omega)$.*

In Theorem 2.26, what is remarkable is that the pointwise convergence of the resolvent operator (which is another way to see the γ-convergence) is actually equivalent to strong convergence. It is due to the compact embedding $H_0^1 \hookrightarrow L^2$. Let us give the proof of this fact for sake of completeness.

Proof of Theorem 2.26 Let us denote by A_n and A the resolvent operators A_{Ω_n} and A_Ω. First of all, we remark that the operators A_n and A have a bounded norm. Indeed, from the variational formulation (2.4), it comes (we set $u_n = A_n(f)$)

$$\int_D |\nabla u_n|^2 \, dx = \int_\Omega f(x) u_n(x) \, dx$$

and, thanks to Poincaré inequality for the left-hand side and Cauchy-Schwarz inequality for the right-hand side, we have

$$\lambda_1(D) \|u_n\|_{L^2}^2 \le \|f\|_{L^2} \|u_n\|_{L^2}$$

which shows that

$$\|A_n\| := \sup_{f \in L^2(D)} \frac{\|u\|_{L^2}}{\|f\|_{L^2}} \le \frac{1}{\lambda_1(D)}. \tag{2.34}$$

Now, we claim that it is possible (for fixed n) to find f^n in the unit ball of $L^2(D)$ achieving the supremum in

$$\sup_{\|f\|_{L^2(D)} \le 1} \|A_n(f) - A(f)\|_{L^2(D)} = \|A_n(f^n) - A(f^n)\|_{L^2(D)}.$$

Indeed, if f_k is a maximizing sequence, it is possible to extract a subsequence which converge weakly to some f^n which also belongs to the unit ball of $L^2(D)$. Since the embedding from $L^2(D)$ into $H^{-1}(D)$ is compact (it is the adjoint of the embedding from $H_0^1(D)$ into $L^2(D)$), and since A_n and A are continuous from $H^{-1}(D)$ to $H_0^1(D)$ the previous equality follows when we let k going to infinity.

Now, let us repeat this method with the sequence f^n: there exists f in the unit ball of $L^2(D)$ such that f^n converge weakly in $L^2(D)$ and strongly in $H^{-1}(D)$ to f. Let us fix an integer n_1 such that for $n \geq n_1$, we have

$$\|f^n - f\|_{H^{-1}(D)} \leq \frac{\varepsilon \lambda_1(D)}{4} \quad \text{and} \quad \|A_n(f) - A(f)\|_{L^2(D)} \leq \frac{\varepsilon}{2},$$

the second inequality coming from the assumption on the γ-convergence of A_n to A. Then

$$\sup_{\|g\|_{L^2(D)} \leq 1} \|A_n(g) - A(g)\|_{L^2(D)} = \|A_n(f^n) - A(f^n)\|_{L^2(D)}$$

$$\leq \|A_n(f) - A(f)\|_{L^2(D)}$$
$$+ \|A_n(f^n - f) - A(f^n - f)\|_{L^2(D)}$$
$$\leq \frac{\varepsilon}{2} + \|A_n - A\|_{\mathscr{L}(H_0^1, H^{-1})} \frac{\varepsilon \lambda_1(D)}{4}$$
$$\leq \frac{\varepsilon}{2} + \frac{2}{\lambda_1(D)} \frac{\varepsilon \lambda_1(D)}{4} = \varepsilon,$$

what proves the desired result.

Now, we give some sufficient conditions ensuring γ-convergence and therefore, convergence of eigenvalues. We refer to [32, chapter 3] for more details.

Theorem 2.28 (Convex Case) *Let B be a fixed compact set in $\mathbb{R}^N$ and Ω_n be a sequence of convex open sets in B which converges, for the Hausdorff distance, to a (convex) set Ω. Then Ω_n γ-converge to Ω and, in particular, for all k fixed, $\lambda_k(\Omega_n) \to \lambda_k(\Omega)$.*

In two-dimension, there is a nice result due to V. Šverak which gives continuity with much weaker assumptions, see [32, 55]. Roughly speaking, it says that if the number of holes in the sequence Ω_n is uniformly bounded and if Ω_n converges for the Hausdorff distance, then there is convergence of eigenvalues. To be more precise, let us introduce, for any open set Ω (whose complementary is denoted by Ω^c):

$$\sharp \Omega^c := \text{number of connected components of } \Omega^c.$$

Theorem 2.29 (Šverak) *Let B be a fixed compact set in $\mathbb{R}^2$ and Ω_n a sequence of open subsets of B. Let p be a given integer and assume that the sets Ω_n satisfy*

Fig. 2.2 The eigenvalues of
the connected set Ω_ε
converge to the eigenvalues of
$\Omega_1 \cup \Omega_2$

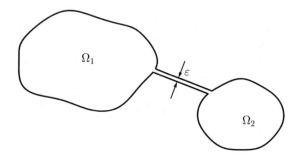

$\sharp\Omega_n^c \leq p$. Then, if the sets Ω_n converge for the Hausdorff distance to a set Ω, they γ-converge to Ω and, in particular, for all k fixed, $\lambda_k(\Omega_n) \to \lambda_k(\Omega)$.

As an example of application (at least in dimension 2), let us now give a continuity result which will be useful in several situations. It is classical, you can find the proof for example in [29]. In particular, it shows that adding a connectedness constraint generally does not change anything in minimization problems.

Theorem 2.30 *Let Ω_1 and Ω_2 be two disjoint open sets in $\mathbb{R}^N$, $N \geq 2$ and let Σ be a segment joining Ω_1 and Ω_2. Let ε be a (small) positive number and let us denote by Ω_ε the open set*

$$\Omega_\varepsilon = \bigcup_{x \in \Sigma} B(x, \varepsilon) \cup \Omega_1 \cup \Omega_2$$

obtained by joining the sets Ω_1 and Ω_2 by a small tube of width ε, see Fig. 2.2. Then, for every integer k,

$$\lambda_k(\Omega_\varepsilon) \to \lambda_k(\Omega_1 \cup \Omega_2) \quad \text{when } \varepsilon \to 0 .$$

$\boxed{\text{end of tool C}}$

We come back to the problem of minimizing λ_1 among polygons with (at most) N sides. To prove existence of a minimizer, the sketch of the proof is as follows:

1. Take a minimizing sequence Ω_n and prove that we can assume it to have uniformly bounded diameter
2. Then, we can assume that Ω_n is in a fixed box D and extract a subsequence which converges for the Hausdorff distance to some open set Ω. Prove that Ω is a polygon with at most N sides.
3. Use Theorem 2.29 to prove continuity of eigenvalues and therefore that Ω is a minimizer.
4. Prove by contradiction that Ω has exactly N sides (if not, improve λ_1 by cutting near a vertex).

2.2.3.1 The Case of Triangles and Quadrilaterals

After the existence result, we would like to identify the minimizer in $\mathscr{P}_N$. According to the Faber-Krahn inequality, it is natural to conjecture that it is the N-regular polygon. Actually, the result is known only for $N = 3$ and $N = 4$:

Theorem 2.31 (Pólya) *The equilateral triangle has the least first eigenvalue among all triangles of given area. The square has the least first eigenvalue among all quadrilaterals of given area.*

Proof The proof relies on the same technique as the Faber-Krahn Theorem with the difference that is now used the Steiner symmetrization. Since this symmetrization has the same properties (2.22) and (2.23) as the Schwarz rearrangement, it is clear that any Steiner symmetrization decreases (or at least do not increase) the first eigenvalue. By a sequence of Steiner symmetrization with respect to the mediator of each side, a given triangle converges to an equilateral one. More precisely, let us denote by h_n and a_n the height and the length of the basis of the triangle T_n that we get at step n and A_n one of the basis angle (see Fig. 2.3). Elementary trigonometry yields

$$\frac{h_n}{a_{n+1}} = \sin A_n , \qquad \frac{h_{n+1}}{a_n} = \sin A_n . \tag{2.35}$$

Let us denote by $x_n := \frac{h_n}{a_n}$. Relations (2.35) reads

$$x_{n+1} = \frac{\sin^2 A_n}{x_n} = \frac{\sin^2(\arctan(2x_n))}{x_n} = \frac{4x_n}{1 + 4x_n^2} .$$

Now, an elementary study of the sequence $x_{n+1} = \frac{4x_n}{1+4x_n^2}$ shows that it converges to the fixed point of $f(x) = \frac{4x}{1+4x^2}$ which is $\frac{\sqrt{3}}{2}$ i.e. the value characteristic of

Fig. 2.3 The triangle T_n and its Steiner symmetrization T_{n+1}

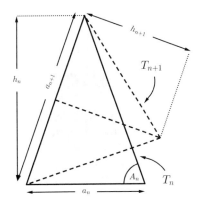

Fig. 2.4 A sequence of 3
Steiner symmetrizations
transforms any quadrilateral
into a rectangle

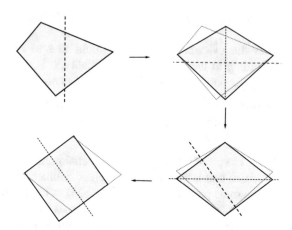

equilateral triangles. Moreover, with the same argument as above (Šverak Theorem), the sequence of triangles γ-converges to the equilateral one, say $\widehat{T}$, so we have proved, if T denotes the triangle we started with:

$$\lambda_1(\widehat{T}) = \lim \lambda_1(T_n) \leq \lambda_1(T) \ .$$

With a more careful study, we can prove that the above inequality is strict if T is not equilateral.

Curiously, the proof is a little bit simpler for quadrilaterals. Indeed a sequence of three Steiner symmetrization allows us to transform any quadrilateral into a rectangle, see Fig. 2.4. Therefore, it suffices to look at the minimization problem among rectangles. But it is elementary to prove that the square is the best rectangle for λ_1, use Sect. 2.1.4.

2.2.3.2 A Challenging Open Problem

Unfortunately, for $N \geq 5$ (pentagons and others), the Steiner symmetrization increases, in general, the number of sides, see Fig. 2.5. This prevents us to use the same technique. So a beautiful (and hard) challenge is to solve the

Open Problem 2.32 *Prove that the regular N-gone has the least first eigenvalue among all the N-gones of given area for $N \geq 5$.*

This conjecture is supported by the classical isoperimetric inequality linking area and length for regular N-gones, see e.g. Theorem 5.1 in Osserman, [48].

Fig. 2.5 The Steiner
symmetrization of a pentagon
has, in general, 6 edges

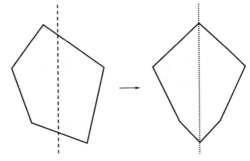

Fig. 2.6 Ω^* solves
Problem (2.36): the free
components of $\partial\Omega^*$ are not
arc of circles

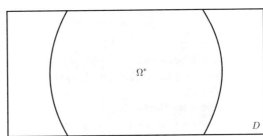

2.2.4   Domains in a Box

Instead of looking at open sets just with a volume constraint, we can consider open sets constrained to lie into a given box D (and also with a given volume). In other words, we could look for the solution of

$$\min\{\lambda_1(\Omega),\ \Omega \subset D,\ |\Omega| = A \text{ (given)}\}. \tag{2.36}$$

According to a classical Theorem of Buttazzo-DalMaso, see [19] the problem (2.36) has always a solution in the class of quasi-open sets. Of course, if the constant A is small enough in such a way that there exists a ball of volume A into the box D, it will provide the solution (since it is the global minimum). Therefore, the interesting case is when any ball of volume A is "too big" to stay into D. Actually, we can prove the following.

Theorem 2.33 *Let $\Omega^* \subset \mathbb{R}^2$ be a minimizer for the problem (2.36). Assume that there is no disk of area A in the box D. Then,*

- *(i) Ω^* touches the boundary of D.*
- *(ii) The free parts of the boundary of Ω^* (i.e. those which are inside D) are analytic.*
- *(iii) The boundary of Ω^* does not contain any arc of circle (Figs. 2.6 and 2.7).*

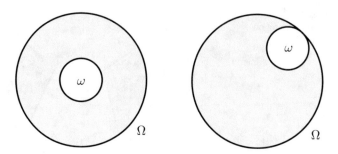

Fig. 2.7 Position of the circular hole which maximizes $\lambda_1(\Omega \setminus \omega)$ (left); one position which minimizes $\lambda_1(\Omega \setminus \omega)$ (right)

Proof We will not prove here point (ii) for which we refer to [11] or [30, chapter 3]. Let us now prove point (iii) (see also [31]). For that purpose, we need the notion of *domain derivative*.

| **Tool D: Domain Derivative** |

Let us define the derivative of an eigenvalue with respect to the domain. We consider an open set Ω and a family of applications $\Phi(t)$ satisfying

$$\Phi : t \in [0, T[\to W^{1,\infty}(\mathbb{R}^N, \mathbb{R}^N) \text{ differentiable at 0 with } \Phi(0) = I, \Phi'(0) = V \tag{2.37}$$

where $W^{1,\infty}(\mathbb{R}^N, \mathbb{R}^N)$ is the set of bounded Lipschitz maps from $\mathbb{R}^N$ into itself, I is the identity and V a vector field. For t small, $\Phi(t)$ is a diffeomorphism. For example, it is classical to choose

$$\Phi(t) = I + tV .$$

Let us denote by $\Omega_t = \Phi(t)(\Omega)$ and by $\lambda_k(t) = \lambda_k(\Omega_t)$ the k-th eigenvalue of the Laplacian on Ω_t. We assume that $\lambda_k(t)$ is simple (for t small) and, since k is fixed in the sequel of this section, we denote by u_t an associated eigenfunction in $H_0^1(\Omega_t)$ with the normalization

$$\int_{\Omega_t} u_t^2(x) \, dx = 1. \tag{2.38}$$

Then, we have

Theorem 2.34 (Hadamard Formulae) *Let Ω be a bounded open set. We assume that $\lambda_k(\Omega)$ is simple. Then, the functions $t \to \lambda_k(t)$, $t \to u_t \in L^2(\mathbb{R}^N)$ are differentiable at $t = 0$ with*

$$\lambda_k'(0) = -\int_\Omega div(|\nabla u|^2 V) \, dx . \tag{2.39}$$

If, moreover, Ω is of class C^2 or if Ω is convex, then

$$\lambda'_k(0) = -\int_{\partial\Omega} \left(\frac{\partial u}{\partial n}\right)^2 V.n \, d\sigma . \tag{2.40}$$

and the derivative u' of u_t is the solution of

$$\begin{cases} -\Delta u' = \lambda_k u' + \lambda'_k u & \text{in } \Omega \\ u' = -\frac{\partial u}{\partial n} V.n & \text{on } \partial\Omega \\ \int_\Omega u \, u' \, d\sigma = 0 . \end{cases} \tag{2.41}$$

Case of Multiple Eigenvalues As it is easily seen, even in the case of a matrix, multiple eigenvalue is no longer differentiable in a classical sense. Therefore, two strategies can be considered.

- We use the sub-differential.
- We look at directional derivatives.

The first strategy is explained, for example, in [20, 22] where the sub-differential is computed. We choose here to present the second strategy, since it will be useful in the sequel. The following result is proved in [54].

Theorem 2.35 (Derivative of a Multiple Eigenvalue) *Let Ω be a bounded open set of class C^2. Assume that $\lambda_k(\Omega)$ is a multiple eigenvalue of order $p \geq 2$. Let us denote by $u_{k_1}, u_{k_2}, \ldots, u_{k_p}$ an orthonormal (for the L^2 scalar product) family of eigenfunctions associated to λ_k. Let $\Phi(t)$ satisfying (2.37) with V **fixed** and $\Omega_t = \Phi(t)(\Omega)$. Then $t \to \lambda_k(\Omega_t)$ has a (directional) derivative at $t = 0$ which is one of the eigenvalues of the $p \times p$ matrix $\mathcal{M}$ defined by:*

$$\mathcal{M} = \left(m_{i,j}\right) \quad \text{with } m_{i,j} = -\int_{\partial\Omega} \left(\frac{\partial u_{k_i}}{\partial n} \frac{\partial u_{k_j}}{\partial n}\right) V.n \, d\sigma \quad i, j = 1, \ldots, p . \tag{2.42}$$

Of course, this Theorem contains the case of a simple eigenvalue, since the matrix $\mathcal{M}$ has then a single entry which is exactly (2.40). When λ_k is a multiple eigenvalue, the directional derivative depends on the choice of the vector field V: changing V make shifting from one eigenvalue to the other of the matrix $\mathcal{M}$. This means that $V \mapsto \lambda'_k(0)$ is no longer a linear form which is another way to express the non-differentiability of $t \to \lambda_k(t)$ at $t = 0$.

In the sequel, we will also use the formulae for the derivative of the volume.

Theorem 2.36 (Derivative of the volume) *Let Ω be a bounded open set and $Vol(t) = |\Omega_t|$ the volume of Ω_t. Then, the function $t \to Vol(t)$ is differentiable at $t = 0$ with*

$$Vol'(0) := \int_\Omega div(V) \, dx . \tag{2.43}$$

Moreover, if Ω is Lipschitz,

$$Vol'(0) := \int_{\partial\Omega} V.n \, d\sigma \, . \tag{2.44}$$

Corollary 2.37 *Let Ω be a convex or C^2 domain in $\mathbb{R}^N$ which minimizes an eigenvalue λ_k among all open sets of given volume. Assume that the eigenvalue $\lambda_k(\Omega)$ is simple. Then, there exists a constant c such that the eigenfunction u_k satisfies*

$$\left| \frac{\partial u_k}{\partial n} \right| = c \quad on \; \partial\Omega \, . \tag{2.45}$$

Indeed, if Ω minimizes λ_k under the constraint $Vol(\Omega) = A$, there exists a Lagrange multiplier C such that $\lambda'_k(0) = CVol'(0)$ which reads

$$- \int_{\partial\Omega} \left(\frac{\partial u_k}{\partial n} \right)^2 V.n \, d\sigma = C \int_{\partial\Omega} V.n \, d\sigma$$

for any vector field V in $W^{1,\infty}(\mathbb{R}^N)$ (we know that the eigenfunction u_k belongs to the Sobolev space $H^2(\Omega)$ by classical regularity results. But this implies $-\left(\frac{\partial u_k}{\partial n} \right)^2 = C$ which gives the desired result with $c = \sqrt{-C}$.

Remark 2.38 It is easy to see that the above constant c cannot be zero (for example using the classical formulae $\lambda_k = \frac{1}{2} \int_{\partial\Omega} \left(\frac{\partial u}{\partial n} \right)^2 X.n \, d\sigma$). Therefore, for a minimizer of λ_k:

- either λ_k is double (see Open Problem 2.64)
- or no nodal line of u_k hits the boundary (otherwise we would have $c = 0$).

$$\boxed{\text{End of Tool D}}$$

We come back to the proof that the optimizer of λ_1 in a box has no arc of circle on its boundary. Let us denote by u the first (normalized) eigenfunction of Ω^*. Let us assume that $\partial\Omega^*$ contains a piece of circle γ. According to Corollary 2.37, Ω^* satisfies the optimality condition

$$\frac{\partial u}{\partial n} = -c \quad on \; \gamma \, . \tag{2.46}$$

We put the origin at the center of the corresponding disk and we introduce the function

$$w(x, y) = x \frac{\partial u}{\partial y} - y \frac{\partial u}{\partial x}.$$

Then, we easily verify that

$$-\Delta w = \lambda_1 w \text{ in } \Omega^*$$
$$w = 0 \text{ on } \gamma$$
$$\frac{\partial w}{\partial n} = 0 \text{ on } \gamma.$$

Now we conclude, using Hölmgren uniqueness theorem, that w must vanish in a neighborhood of γ, so in the whole domain by analyticity. Now, it is classical that $w = 0$ imply that u is radially symmetric in Ω^*. Indeed, in polar coordinates, $w = 0$ implies $\frac{\partial u}{\partial \theta} = 0$. Therefore Ω^* is a disk. At last, to prove (i), we can use the optimality condition to see that the eigenfunction should satisfy $|\nabla u| = cst$ on the whole boundary. Then, we can conclude that Ω^* should be a ball, thanks to a classical result by J. Serrin on such overdetermined problems.

Remark 2.39 The previous theorem partly generalizes in higher dimension. Actually, points (i) and (iii) can be proved exactly in the same way. For example, for point (iii) we use the functions

$$w_{i,j} := x_i \frac{\partial u}{\partial x_j} - x_j \frac{\partial u}{\partial x_i}, \quad i, j = 1, \ldots, N$$

instead of w and we prove that all these functions $w_{i,j}$ vanish in Ω^* which implies that Ω^* is a ball. The regularity is not completely known. In [12] it is proved that if the box D is connected, then the reduced boundary of the optimal domain inside D is an analytic hypersurface and its complement has $N - 1$-Hausdorff measure zero. We refer to [30, Chapter 3] for a good survey on these questions of regularity.

Open Problem 2.40 *Let $\Omega^* \subset \mathbb{R}^N$ be a minimizer for the problem (2.36). Prove that $\partial \Omega^* \cap D$ is analytic in dimension $N \leq 7$.*

We can also consider open questions related to the geometry of the minimizer:

Open Problem 2.41 *Let $\Omega^* \subset \mathbb{R}^N$ be a minimizer for the problem (2.36). Prove that D convex (resp. starshaped) implies that Ω^* is convex (resp. starshaped).*

2.2.5 Multi-Connected Domains

This section could also be entitled "How to place an obstacle" (see [28]). Let Ω be a bounded open set which is fixed in all that section. Two different problems can be considered in that context:

(P1) **location of a given obstacle**: we consider a fixed obstacle, for example a given ball B and we look for the position of this ball into Ω in order to minimize or maximize the first eigenvalue of $\lambda_1(\Omega \setminus B)$,

(P2)   **shape and location of an obstacle**: now we look for both the shape and position of an obstacle K into Ω with a volume or perimeter constraint in order to minimize or maximize the first eigenvalue of $\lambda_1(\Omega \setminus K)$.

2.2.5.1   Optimizing with a Given Obstacle

Obviously, Problem P1 is simpler. The particular case where Ω is itself a ball has been studied several times, starting by J. Hersch in [34] for the two-dimensional case and extended to the N-dimensional case by several authors: M. Ashbaugh and T. Chatelain in 1997 (private communication), Ramm and Shivakumar see [53], E. Harrell, P. Kröger and K. Kurata in [28], Kesavan, see [37]. They also proved that $\lambda_1(B_1 \setminus B_0)$ is minimum when B_0 touches the boundary of B_1. Let us state the result in its more general form (see [28])

Theorem 2.42 (Harrell-Kröger-Kurata) *Let Ω be a bounded convex domain in $\mathbb{R}^N$ and B a ball contained in Ω. Assume that Ω is symmetric with respect to some hyperplane H. We are interested in the position of B which maximizes or minimizes the first Dirichlet eigenvalue $\lambda_1(\Omega \setminus B)$. Then*

- *at the minimizing position B touches the boundary of Ω*
- *at the maximizing position B is centred on H.*

Proof First of all, existence of a minimizing or maximizing position is not difficult to get here. Indeed the only variable is the center of the ball (which stay in a compact set) and the continuity of the eigenvalue w.r.t. the center of the ball is classical.

Let us assume that the ball is not in one of the positions described above: it does not touch the boundary and it is not centred on H. The result will be established if we prove that $\lambda_1(\Omega \setminus B)$ decreases when B moves away from H. Without loss of generality, we can assume that e_1 is the normal direction to H. If $B(X_0, \rho)$ is the initial position of the ball, we look at the function

$$t \mapsto \lambda(t) := \lambda_1(\Omega \setminus B(X_0 + te_1, \rho)).$$

Using the Hadamard formulae (2.40), we can see that $t \mapsto \lambda(t)$ is derivable at 0 and

$$\lambda'(0) = -\int_{\partial B} \left(\frac{\partial u}{\partial n}\right)^2 n_1. \tag{2.47}$$

where u is the normalized eigenfunction associated to $\lambda_1(\Omega \setminus B(X_0, \rho))$ and n_1 the first coordinate of the exterior normal vector. So it suffices to prove that $\lambda'(0) < 0$. For that purpose, we use some kind of moving plane method. Let $T : \{x = x_1\}$ denotes an hyperplane parallel to H passing by X_0 and $\omega^+ = \{x \in \Omega \setminus B; \; x > x_1\}$ (see Fig. 2.8). By assumption on Ω, the reflection of ω^+ through T is strictly included in $\Omega \setminus B$ (this is the crucial point). For any $X \in \omega^+$, we denote by X' its reflection through T. We introduce $w(X) = u(X) - u(X')$ defined

Fig. 2.8 The moving plane method applied to prove that the derivative of λ_1 is negative

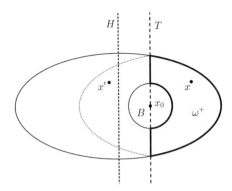

on ω^+. By construction, this function w vanishes on T and on $\partial B \cap \partial \omega^+$ and $w < 0$ on $\partial \Omega \cap \partial \omega^+$. Moreover, $-\Delta w = \lambda_1 w$ where $\lambda_1 = \lambda_1(\Omega \setminus B)$. Since $\lambda_1(\Omega \setminus B) < \lambda_1(\omega^*)$ (by monotonicity property of eigenvalues), the generalized maximum principle applies and $w < 0$ in ω^+. Moreover, since w attained its maximum at $X \in \partial B \cap \partial \omega^+$ (which is C^2 except at points of $T \cap \partial B \cap \partial \omega^+$), the Hopf's boundary point lemma applies and:

$$\frac{\partial w(X)}{\partial n} = \frac{\partial u(X)}{\partial n} - \frac{\partial u(X')}{\partial n} > 0 \, .$$

This property with $\frac{\partial u(X)}{\partial n} < 0$ on ∂B implies

$$\left(\frac{\partial u(X)}{\partial n} \right)^2 < \left(\frac{\partial u(X')}{\partial n} \right)^2$$

which gives $\lambda'(0) < 0$ thanks to (2.47) and a decomposition of the integral in a sum of integrals over the two hemispheres.

With more assumptions and in two dimensions, one can state a more precise result:

Theorem 2.43 (Harrel-Kröger-Kurata) *Let Ω be a C^2 convex domain in $\mathbb{R}^2$. Assume that Ω is symmetric with respect to two perpendicular lines, say Ox and Oy. Assume, moreover, that in each quadrant of the plane, the curvature of the boundary of Ω is monotonic as a function of x.*

Now, let B be a ball of radius ρ with ρ less than the maximum of the curvature of $\partial \Omega$ (attained at a point which is called a vertex of Ω). Then

- $\lambda_1(\Omega \setminus B)$ *is minimum when B is in contact with a vertex*
- $\lambda_1(\Omega \setminus B)$ *is maximum when B is centered at the origin.*

For the proof, see [28].

An interesting open question is to generalize the previous theorems of Harrel-Kröger-Kurata:

Open Problem 2.44 *Let Ω be a fixed domain in $\mathbb{R}^N$ and B_0 a ball of fixed radius. Prove that $\lambda_1(\Omega \setminus B_0)$ is minimal when B_0 touches the boundary of Ω (where?) and is maximum when B_0 is centred at a particular point of Ω (at what point?).*

Actually, for the maximization problem, it seems that the optimal center of B_0 depends on the radius and is not fixed (apart in the case of symmetries). When the radius of B_0 goes to zero, classical asymptotic formulae for eigenvalues of domains with small holes, see (2.21) and the review paper [26], lead one to think that the ball must be located at a maximal point of the first eigenfunction of the domain without holes.

2.2.5.2 Finding the Shape and the Location of the Obstacle

In that case, we are looking for some compact $K \subset \Omega$ in order to minimize or maximize $\lambda_1(\Omega \setminus K)$. The first step is to choose the good constraint:

Problem 2.45 (Minimizing the First Eigenvalue with an Obstacle of Fixed Area) For a fixed $A \in (0, |\Omega|)$, consider the problem

$$\min\{\lambda_1(\Omega \setminus K) : K \subset \overline{\Omega}, K \text{ closed}, |K| = A\}. \tag{2.48}$$

This problem is strictly related to the minimization of the first eigenvalue for domains contained in a *box*, which has been analysed in Sect. 2.2.4, see also [29, Section 3.4]. Indeed, setting $C = \Omega \setminus K$, problem (2.48) becomes equivalent to the minimization of $\lambda_1(C)$ among open sets $C \subseteq \Omega$ of area $|\Omega| - A$, where Ω plays the role of a box containing all possible competitors.

Problem 2.46 (Maximizing the First Eigenvalue with an Obstacle of Fixed Area) The corresponding *maximization* problem of (2.48) has no solution. Indeed one can construct a family of closed sets K_n of fixed area A so as the first eigenvalue $\lambda_1(\Omega \setminus K_n) \uparrow \infty$ as $n \to \infty$ (for instance take K_n as the union of a fixed closed set of area A and a curve γ_n filling Ω as n increases, see [56] where the limit distribution in $\overline{\Omega}$ of these sets is studied in detail). To have existence of a maximizer one needs to require stronger geometrical and topological constraints in the class of admissible obstacles, preventing maximizing sequences to spread out over Ω (notice that connectedness is still not sufficient). Therefore, we are led to formulate the following problem: for a fixed $A \in (0, |\Omega|)$

$$\max\{\lambda_1(\Omega \setminus K) : K \subset \overline{\Omega}, K \text{ closed and convex}, |K| = A\}. \tag{2.49}$$

The existence of a maximizer in the class of convex sets is straightforward (see [15, 29]). Moreover, as convexity seems necessary for the existence, it is natural to expect every solution in (2.49) to *saturate* the convexity constraint, in the sense that its boundary should contain non-strictly convex parts. In particular it could be interesting to know whether this problem has only polygonal sets as solutions, see [42, 43] for results in this direction for shape optimization problems with convexity constraints.

Problem 2.47 (Minimizing the First Eigenvalue with an Obstacle of Fixed Perimeter)   Also the corresponding minimization problem of (2.48) with the constraint on the perimeter (i.e., one dimensional Hausdorff measure of the boundary) has no solutions, since one can construct a family of closed sets K_n of fixed perimeter L so as the first eigenvalue $\lambda_1(\Omega \setminus K_n) \downarrow \lambda_1(\Omega)$ as $n \to \infty$ (for instance take K_n so as its boundary ∂K_n is a closed curve with fixed perimeter converging, as n increases, to a subset Γ of $\partial \Omega$). Therefore, as in Problem 2.46 we are forced to restrict the class of admissible obstacles to convex sets and we formulate the following problem: for a fixed $L \in (0, \mathcal{H}^1(\partial \Omega))$

$$\min\{\lambda_1(\Omega \setminus K) : K \subset \overline{\Omega}, \ K \text{ closed and convex}, \ \mathcal{H}^1(\partial K) = L\}, \qquad (2.50)$$

where $\mathcal{H}^1$ denotes the one dimensional Hausdorff measure (if Ω is not regular $\mathcal{H}^1(\partial \Omega) = \infty$ by convention). Now, even in this smaller class of admissible obstacles the existence question is not so clear. For example, if the boundary of Ω contains a segment and if L is small enough (smaller than twice the length of the segment), it is still possible to build a minimizing sequences of convex domains K_n approaching the boundary of Ω and so that $\lambda_1(\Omega \setminus K_n) \downarrow \lambda_1(\Omega)$. On the other hand, if L is large enough, existence of a minimizer is straightforward since minimizing sequences will not be able to converge to a segment. In any case, one expects solutions of (2.50) touching the boundary $\partial \Omega$.

Problem 2.48 (Maximizing the First Eigenvalue with an Obstacle of Fixed Perimeter)   Now, assuming moreover connectedness of the obstacle K, the maximization problem is well posed in the plane and has been widely studied in the recent paper [33]. The first issue is to introduce a good notion of perimeter. Indeed, since objects of positive capacity but zero Lebesgue measure influence the first eigenvalue but are not seen by the classical perimeter (as defined by De Giorgi), we need to choose another notion of perimeter more sensitive to one-dimensional objects. Moreover, it is natural to ask that this notion of perimeter

1. coincides with the classical notion of perimeter on *regular* sets;
2. is continuous (or at least lower semicontinuous) for Hausdorff convergence. In particular, this perimeter measures twice the length of one dimensional objects.

To these purposes it is possible to work with the *outer Minkowski content* (see for instance [2] where this quantity is studied in detail): for a closed set K in $\overline{\Omega}$, whenever the limit exits, we define

$$\mathcal{M}(K) := \lim_{\epsilon \to 0} \frac{|K^\epsilon \setminus K|}{\epsilon}, \qquad (2.51)$$

where $K^\epsilon := \{x \in \mathbb{R}^2 : d(x, K) \leq \epsilon\}$ is the tubular neighborhood of K through the distance function $d(\cdot, K)$ to K. According to this definition, the outer Minkowski content of any closed set in $\overline{\Omega}$ with Lipschitz boundary coincides with the classical perimeter $\mathscr{P}$ (in the sense of De Giorgi) and with the Hausdorff measure of the boundary.

Now, for a domain $\Omega \subset \mathbb{R}^2$ with Lipschitz boundary and for a fixed $L \in (0, \mathscr{H}^1(\partial\Omega))$ we study the maximization problem

$$\max\{\lambda_1(\Omega \setminus K) : K \subseteq \overline{\Omega}, \ K \text{ compact and connected}, \mathscr{M}(K) \leq L\}, \qquad (2.52)$$

Here connectedness of the admissible obstacles combined with the perimeter constraint, prevents maximizing sequences to spread out over Ω and it is sufficient for the existence a solution. For that problem, we can prove, see [33]:

Theorem 2.49 (Henrot-Zucco) *There exists a maximizer K^* of problem* (2.52). *It has the following properties*

(i) K^* *is locally convex in* Ω. *Moreover, if* Ω *is convex then* K^* *is convex.*
(ii) *The perimeter constraint is saturated, namely* $\mathscr{M}(K^*) = L$.
(iii) K^* *is locally either a segment or regular* (C^∞) *inside* Ω.
(iv) *When* Ω *is a disk, the optimal obstacle* K^* *is a centered disk.*

We also studied the case where Ω is an annulus and prove breaking of symmetry for some values of the perimeter constraint L.

2.3 The Second Eigenvalue

2.3.1 Minimizing λ_2

We are now interested in minimizing the second eigenvalue of the Laplacian-Dirichlet among open sets of given volume. As we are going to see, the minimizer is no longer one ball, but two. This result is sometimes attributed to P. Szego (G. Szegö's son), cf [51], but actually it was already contained (more or less explicitly) in one of Krahn's paper, [40]. It has also been rediscovered independently by a Japanese mathematician, Imsik Hong in the 1950s, see [35] (M. Ashbaugh kindly draws my attention to this reference).

Theorem 2.50 (Hong-Krahn-Szego) *The minimum of $\lambda_2(\Omega)$ among bounded open sets of $\mathbb{R}^N$ with given volume is achieved by the union of two identical balls.*

Proof Let Ω be any bounded connected open set (if Ω is not connected, see below). Let us denote by Ω_+ and Ω_- its nodal domains, see Sect. 2.1.5.2. We already know (Proposition 2.12) that $\lambda_2(\Omega)$ is the first eigenvalue for Ω_+ and Ω_-:

$$\lambda_1(\Omega_+) = \lambda_1(\Omega_-) = \lambda_2(\Omega). \qquad (2.53)$$

We now introduce Ω_+^* and Ω_-^* the balls of same volume as Ω_+ and Ω_- respectively. According to the Faber-Krahn inequality

$$\lambda_1(\Omega_+^*) \leq \lambda_1(\Omega_+), \qquad \lambda_1(\Omega_-^*) \leq \lambda_1(\Omega_-). \qquad (2.54)$$

Let us introduce a new open set $\widetilde{\Omega}$ defined as

$$\widetilde{\Omega} = \Omega_+^* \cup \Omega_-^*.$$

Since $\widetilde{\Omega}$ is disconnected, we obtain its eigenvalues by gathering and reordering the eigenvalues of Ω_+^* and Ω_-^*. Therefore,

$$\lambda_2(\widetilde{\Omega}) \leq \max(\lambda_1(\Omega_+^*), \lambda_1(\Omega_-^*)).$$

According to (2.53), (2.54) we have

$$\lambda_2(\widetilde{\Omega}) \leq \max(\lambda_1(\Omega_+), \lambda_1(\Omega_-)) = \lambda_2(\Omega).$$

If Ω would not be connected at the beginning, $\Omega = \Omega_1 \cup \Omega_2$, the proof would be the same by applying the argument to Ω_1 and Ω_2 instead of Ω_+ and Ω_-. This shows that, in any case, the minimum of λ_2 is to be searched among the union of balls. But, if the two balls would have different radii, we would decrease the second eigenvalue by shrinking the largest one and dilating the smaller one (without changing the total volume). Therefore, the minimum is achieved by the union of two identical balls.

Remark 2.51 If we are disappointed with the solution of our minimization problem, since it is not connected, we could think to look at the following problem:

$$\min\{\lambda_2(\Omega), \ \Omega \text{ connected open subset of } \mathbb{R}^N, \ |\Omega| = c\}. \qquad (2.55)$$

Unfortunately, this problem has no solution: indeed, let us consider the set Ω_ε obtained by joining the two identical balls $B_1 \cup B_2$ (each of volume $c/2$) by a thin pipe of width ε then, according to Theorem 2.30, $\lambda_2(\Omega_\varepsilon) \to \lambda_2(B_1 \cup B_2)$. Of course, Ω_ε does not satisfy the volume constraint. But, if we remember that it is equivalent to minimize the product $\lambda_2(\Omega)|\Omega|^{2/N}$ (see 2.11), it is clear that

$$\lambda_2(\Omega_\varepsilon)|\Omega_\varepsilon|^{2/N} \to \lambda_2(B_1 \cup B_2)|B_1 \cup B_2|^{2/N}$$

where the right-hand side is the optimal value without connectedness assumption and therefore, Ω_ε is a minimizing sequence.

As we did previously for the first eigenvalue and the Faber-Krahn inequality, let us now give a quantitative version of this result. It has been done by L. Brasco and A. Pratelli in [8]. To my knowledge, it is the first time that such a quantitative inequality has been obtained for a domain which is different from the ball. First the authors have introduced a suitable variant of the Fraenkel asymmetry. This is the

Fraenkel 2-*asymmetry*, which measures the L^1 distance of a set from the collection of disjoint pairs of equal balls. It is given by

$$\mathscr{A}_2(\Omega) := \inf \left\{ \frac{|\Omega \Delta (B_+ \cup B_-)|}{|\Omega|} : B_+, B_- \text{ balls s.t.} \begin{array}{l} B_+ \cap B_- = \emptyset, \\ |B_+| = |B_-| = |\Omega|/2 \end{array} \right\}.$$

We then have the following quantitative version of the Hong-Krahn-Szego inequality.

Theorem 2.52 (Brasco-Pratelli) *Let* $\Omega \subset \mathbb{R}^N$ *be an open set with finite measure. Then*

$$|\Omega|^{2/N} \lambda_2(\Omega) - 2^{2/N} |B|^{2/N} \lambda_1(B) \geq \frac{1}{C_N} \mathscr{A}_2(\Omega)^{N+1}, \tag{2.56}$$

for a constant $C_N > 0$ *depending on the dimension* N *only.*

2.3.2 A Convexity Constraint

Now, the problem becomes again interesting if we ask the question to find the **convex** domain, of given volume, which minimizes λ_2. Existence of a minimizer Ω^* follows from Theorems 2.24 and 2.28. Of course, the difficulty is to find it! For sake of simplicity, we restrict us here to the two-dimensional case.

In a paper of 1973 [57], Troesch did some numerical experiments which led him to conjecture that the solution was *a stadium*: the convex hull of two identical tangent disks. It is actually the convex domain which is the closest to the solution without convexity constraint. In [31], we refute this conjecture, see Theorem 2.55 below. Nevertheless, the minimizer looks like very much a stadium!

2.3.2.1 Optimality Conditions

The first step is to prove that the second eigenvalue is simple at the optimal domain, for which we refer to [29, Theorem 2.5.10]

Theorem 2.53 *Let* Ω^* *be a convex domain minimizing the second eigenvalue* λ_2 *(among convex domains of given volume). Assume that* Ω^* *is of class* $C^{1,1}$. *Then* $\lambda_2(\Omega^*)$ *is simple.*

Now we can derive some optimality condition like $|\nabla u_2| = c$ on $\partial \Omega^*$ as established in Corollary 2.37 for the general case. Now, it is not so obvious here due to the convexity constraint. The difficulty is to take care of the convexity constraint when deforming the original domain Ω^* by a vector field V. Indeed, if we perform a small deformation of a strictly convex part of the boundary of Ω^*, this part will not remain necessarily convex, but we can use the fact that the difference between the deformed boundary ant its convex hull is so small, that for first order terms, the formulae of derivative still holds. On the contrary, for segments included in the boundary, it is no longer true. Therefore, we need to make a distinction between the strictly convex parts of the boundary and the segments included in the boundary. Let us mention that the first part of the following theorem holds for any dimension while the second part is strictly two-dimensional.

Theorem 2.54 (Henrot-Oudet)

- *There exists a positive constant α such that the gradient of the eigenfunction u is constant on every strictly convex part of the boundary of Ω^*:*

$$\text{for every } \gamma, \text{ strictly convex part of } \partial\Omega^*, \forall x \in \gamma \quad |\nabla u(x)| = \alpha. \quad (2.57)$$

  Moreover α is given by

$$\alpha^2 = \frac{\lambda_2}{|\Omega^*|}. \quad (2.58)$$

- *If Σ is a segment included in the boundary of Ω^*, let t, $t \in [a, b]$, a parametrization of the segment (the boundary is assumed to be oriented in the clockwise sense), then there exists a non negative function w defined on $[a, b]$ with triple roots at a and b, such that*

$$|\nabla u(t)|^2 = \alpha^2 + w''(t). \quad (2.59)$$

We refer to [29, Theorem 4.2.2] for a detailed proof.

Corollary 2.55 *The optimal domain Ω^* is not a stadium.*

Indeed, the proof follows exactly the same line as the proof of (iii) of Theorem 2.33.

2.3.2.2 Regularity of the Optimal Domain

The C^1 regularity is not hard to prove, see e.g. [13] or [31] for details (we just prove that we can improve the product of the second eigenvalue and the area by cutting a possible corner). The more precise result below is due to J. Lamboley in [41].

Theorem 2.56 (Lamboley) *The minimizer Ω^* is globally $C^{1,\frac{1}{2}}$ and no more. The strictly convex parts are C^∞.*

2.4 The Other Dirichlet Eigenvalues

2.4.1 Existence

In this section we investigate the existence of a domain which minimizes $\lambda_k(\Omega)$ over sets of fixed volume in $\mathbb{R}^N$ for $k \geq 3$. The "bounded" case, where Ω is supposed to lie in a fixed box D have been first considered by G. Buttazzo and G. Dal Maso in 1993, see [19]. More precisely, the Buttazzo-Dal Maso's Theorem states that, for any bounded open set $D \subset \mathbb{R}^N$, the following problem

$$\min_{A \subset D, |A|=c} \lambda_k(A) \tag{2.60}$$

has a solution. Of course, the minimizer depends a priori on the choice of the design region D.

In order to prove the existence of a *global* minimizer, the main difficulty is the passage from the bounded set D to $\mathbb{R}^N$. The main reason for which the previous result fails if $D = \mathbb{R}^N$ is the lack of compactness of the injection $H^1(\mathbb{R}^N) \hookrightarrow L^2(\mathbb{R}^N)$. Thus we had to wait for almost 20 years to get the final existence result:

Theorem 2.57 (Bucur, Mazzoleni-Pratelli) *The problem*

$$\min_{\Omega \subset \mathbb{R}^N, |\Omega|=c} \lambda_k(\Omega) \tag{2.61}$$

has a solution. This one is bounded and has finite perimeter.

The two papers [14] and [45] appeared independently and the same week on Arxiv! In some sense, it is the story of Faber and Krahn which is repeated but with modern tools: Internet and Arxiv versus postal mail and classical submission. These two papers use a completely different strategy. In [14], the notion of *shape subsolution* for the torsion energy is introduced, and it is proved that every such subsolution has to be bounded and has finite perimeter. A second argument, shows that minimizers for (2.61) are shape subsolutions, so they are bounded. The author finishes the proof by using a concentration-compactness argument, like in [17] where it has already be used for the third eigenvalue. The approach of [45] is different: a surgery result proves that some parts (like long and tiny tentacles), can be cut out from every set such that, after small modifications and rescaling, the new set has a diameter uniformly bounded and its first k eigenvalues are smaller. In this way, the existence problem in $\mathbb{R}^N$ can be reduced to the local case of Buttazzo and Dal Maso.

2.4.2 Connectedness of Minimizers

In this section, we denote by Ω_n^* an (quasi-) open set which minimizes λ_n (among open sets of volume 1) and $\lambda_n^* = \lambda_n(\Omega_n^*)$ the minimal value of λ_n. We will also denote by $t\Omega$ the image of Ω by an homothety (or dilation) of ratio t. The following result is, in some sense, a generalization of Hong-Krahn-Szegö's Theorem 2.50. Roughly speaking, it asserts that if a minimizer of λ_n is not connected, each connected component is a minimizer for a lower eigenvalue. These results come from [58] and turn out to be very useful even from a numerical point of view, in particular to rule out some competitors.

Theorem 2.58 (Wolf-Keller) *Let us assume that Ω_n^* is the union of (at least) two disjoints sets, each of them with positive measure. Then*

$$\left(\lambda_n^*\right)^{N/2} = \left(\lambda_i^*\right)^{N/2} + \left(\lambda_{n-i}^*\right)^{N/2} = \min_{1 \le j \le (n-1)/2} \left(\left(\lambda_j^*\right)^{N/2} + \left(\lambda_{n-j}^*\right)^{N/2}\right) \quad (2.62)$$

where, in the previous equality, i is a value of $j \le (n-1)/2$ which minimizes the sum $\left(\lambda_j^*\right)^{N/2} + \left(\lambda_{n-j}^*\right)^{N/2}$. Moreover,

$$\Omega_n^* = \left[\left(\frac{\lambda_i^*}{\lambda_n^*}\right)^{1/2} \Omega_i^*\right] \bigcup \left[\left(\frac{\lambda_{n-i}^*}{\lambda_n^*}\right)^{1/2} \Omega_{n-i}^*\right] \quad \text{(disjoint union)}. \quad (2.63)$$

Proof Let us write $\Omega_n^* = \Omega_1 \cup \Omega_2$ (disjoint union) with $|\Omega_1| > 0$, $|\Omega_2| > 0$ and $|\Omega_1| + |\Omega_2| = 1$. Let u_n^* be an eigenfunction of the Laplacian-Dirichlet on Ω_n^*, corresponding to the eigenvalue λ_n^*. Then u_n^* is not zero on one of the components of Ω_n^*, for example Ω_1. In particular λ_n^* is an eigenvalue (see (2.7)) of Ω_1: $\lambda_n^* = \lambda_i(\Omega_1)$ for some integer $i \le n$ and we denote precisely by i the largest one. If we had $i = n$, we could decrease λ_n^* by enlarging Ω_1 contradicting the minimality of λ_n^*, so $i \le n - 1$. Since λ_n^* is the n-th eigenvalue of Ω_n^*, we can count at least $n - i$ eigenvalues of Ω_2 which are smaller than λ_n^*. It means that $\lambda_{n-i}(\Omega_2) \le \lambda_n^*$. Moreover, if we would have $\lambda_{n-i}(\Omega_2) < \lambda_n^*$, we could decrease $\lambda_n^* = \max\{\lambda_i(\Omega_1), \lambda_{n-i}(\Omega_2)\}$ by enlarging Ω_1 and shrinking Ω_2 (keeping the total volume equal to one) which would again contradict the minimality of λ_n^*. So, finally $\lambda_{n-i}(\Omega_2) = \lambda_i(\Omega_1) = \lambda_n^*$.

Now, we still get a minimum for λ_n^* by replacing Ω_1 by $|\Omega_1|^{1/N} \Omega_i^*$ (which has same volume and better λ_i) and by replacing Ω_2 by $|\Omega_2|^{1/N} \Omega_{n-i}^*$. Consequently, we have

$$\lambda_i(\Omega_1) = |\Omega_1|^{-2/N} \lambda_n^* = \lambda_n^* = |\Omega_2|^{-2/N} \lambda_{n-i}^* = \lambda_{n-i}(\Omega_2).$$

Finally, the constraint $|\Omega_1| + |\Omega_2| = 1$ yields $\left(\lambda_n^*\right)^{N/2} = \left(\lambda_i^*\right)^{N/2} + \left(\lambda_{n-i}^*\right)^{N/2}$.

Let us now consider the set $\widetilde{\Omega}_j$ defined for $j = 1, \ldots, n-1$ by

$$
\widetilde{\Omega}_j = \left[\left(\frac{\left(\lambda_j^* \right)^{N/2}}{\left(\lambda_j^* \right)^{N/2} + \left(\lambda_{n-j}^* \right)^{N/2}} \right)^{1/N} \Omega_j^* \right] \bigcup \left[\left(\frac{\left(\lambda_{n-j}^* \right)^{N/2}}{\left(\lambda_j^* \right)^{N/2} + \left(\lambda_{n-j}^* \right)^{N/2}} \right)^{1/N} \Omega_{n-j}^* \right].
$$
$$(2.64)$$

Each $\widetilde{\Omega}_j$ has a volume equal to one, the j-th eigenvalue of its first component and the $(n-j)$-th eigenvalue of its second component are equal to

$$
\left(\left(\lambda_j^* \right)^{N/2} + \left(\lambda_{n-j}^* \right)^{N/2} \right)^{2/N}.
$$

It follows that $\lambda_n(\widetilde{\Omega}_j)$ is also given by this common value. Since $\lambda_n^* \leq \lambda_n(\widetilde{\Omega}_j)$ and $\lambda_n^* = \lambda_n(\widetilde{\Omega}_i)$ for some index i, λ_n^* is the minimum value of $\lambda_n(\widetilde{\Omega}_j)$. Moreover, $\widetilde{\Omega}_i$ is optimal for any index i which realizes the minimum in (2.62). This finishes the proof.

Up to now, the value of λ_n^* is not known but for $n = 1$ or 2. The previous theorem has an important consequence for the optimal domain Ω_3^* in dimension 2 or 3.

Corollary 2.59 (Wolff-Keller) *Let Ω_3^* be an open set minimizing λ_3 (i.e. solution of problem (2.60) with $c = 1$) in dimension 2 or 3. Then, Ω_3^* is connected.*

Proof Assume that Ω_3^* is not connected. Then, according to Theorem 2.58, we should have $\lambda_3^* = \lambda_1^* + \lambda_2^*$ ($i = 1$ is the only possible value here). Let us explicit these values first in dimension 2. From Theorem 2.14 $\lambda_1^* = \pi j_{0,1}^2 \simeq 18.168$ (we recall that $j_{0,1}$ is the first zero of the Bessel function J_0 and the radius of the disk of area 1 is $R_0 = \frac{1}{\sqrt{\pi}}$, see (2.16)) while, according to Theorem 2.50, $\lambda_2^* = 2\lambda_1^* \simeq 36.336$. Therefore $\lambda_1^* + \lambda_2^* \simeq 54.504$. But since λ_3^* is, by definition, lower or equal to the third eigenvalue of the unit disk $\lambda_3(D_1) = \pi j_{1,1}^2 \simeq 46.125$, we see that it cannot be equal to $\lambda_1^* + \lambda_2^*$.

In dimension 3, the situation is exactly the same. We use the values of the eigenvalues for a ball given in (2.17). Since the ball must have for volume 1, its radius is here $R_0 = \left(\frac{3}{4\pi} \right)^{1/3}$, then

$$
\lambda_1^* = \frac{(4\pi)^{2/3} j_{1/2,1}^2}{3^{2/3}} \simeq 25.646 \qquad \lambda_2^* = 2^{2/3} \lambda_1^* \simeq 40.711
$$

and

$$
\left(\lambda_3(B_{R_0}) \right)^{3/2} = \left(\frac{j_{N/2,1}^2}{R_0^2} \right)^{3/2} \simeq 380.029 < \left(\lambda_1^* \right)^{3/2} + \left(\lambda_2^* \right)^{3/2} \simeq 389.636
$$

with the same conclusion.

Remark 2.60 In dimension 4 and higher, this computation does not give anything. For example, we get in dimension 4:

$$R_0 = \frac{2^{1/4}}{\sqrt{\pi}} \quad \lambda_1^* = \frac{j_{1,1}^2}{R_0^2} \simeq 32.615 \quad \lambda_2^* = \sqrt{2}\lambda_1^* \simeq 46.125$$

while

$$\left(\lambda_3(B_{R_0})\right)^2 = \left(\frac{j_{2,1}^2}{R_0^2}\right)^2 \simeq 3432.67 > \left(\lambda_1^*\right)^2 + \left(\lambda_2^*\right)^2 \simeq 3191.25$$

which cannot help to conclude. Actually, the conjecture is that in dimension $N \geq 4$, Ω_3^* is not connected and, more precisely, that it is given by the union of three balls.

Open Problem 2.61 *Prove that the optimal domain Ω_3^* is the disk in dimension 2, and the union of three identical balls in dimension $N \geq 4$.*

Remark 2.62 In dimension 2, numerical computations show that the optimal domain Ω_n^* is generally connected but may be disconnected, this is the case for $n = 2, 4$, see [3, 49] and figures below.

In Fig. 2.9 we plot the minimizers of the first 12 Dirichlet eigenvalues obtained independently by E. Oudet and P. Antunes and P. Freitas (those pictures are taken from [3], see also [49] and [30, chapter 11]).

Remark 2.63 As you can see on Fig. 2.9, for the 4th first eigenvalues, the optimal domain seems to be a disk or union of two disks (as mentioned above, it has be proved only for $k = 1, 2$). In a recent paper, see [6], Amandine Berger proves that these are the only cases, in dimension 2, for which disks or union of disks can be the optimal domain.

Table 2.1 shows the optimal Dirichlet eigenvalues, together with the corresponding multiplicity of each optimal eigenvalue. As you can see, except for $k = 1$, for all the optimal domains the corresponding eigenvalue is multiple (with a multiplicity which seems to increase).

Open Problem 2.64 *Prove that the eigenvalue for the optimal domain Ω_n^*, $n \geq 2$ is always multiple: $\lambda_{n-1}(\Omega_n^*) = \lambda_n(\Omega_n^*)$.*

Next, we present some numerical results with three-dimensional domains. In Fig. 2.10 we plot the 3D minimizers of the first 10 Dirichlet eigenvalues. As you can see, the ball is no longer the optimal domain for λ_3 in $\mathbb{R}^3$ but it seems to be the optimal domain for λ_4. This leads to the following open problem: A generalization of the open problem for λ_3 in the plane:

Open Problem 2.65 *Prove that the N-ball minimizes λ_{N+1} among sets of given volume in $\mathbb{R}^N$.*

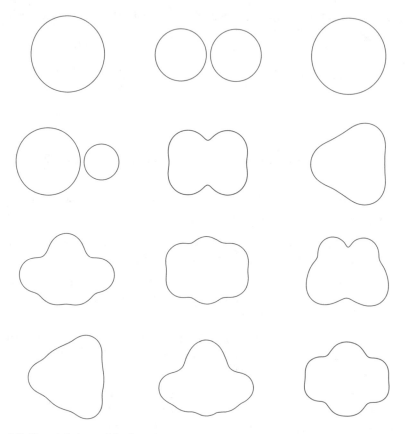

Fig. 2.9 The minimizers of the first 12 Dirichlet eigenvalues (by courtesy of P. Antunes)

2.4.3 Other Geometric Constraints

Instead of looking at the minimization problem with a volume constraint, it is possible to consider other geometric constraints like a constraint on the perimeter or on the diameter.

2.4.3.1 Perimeter Constraint

Considering a perimeter constraint, the situation is easier in dimension two. Indeed, since the convex hull of any domain has a smaller perimeter (what is not true in higher dimension), it is easy to see that we can restrict the study to plane convex domains. For that particular case, the following results are proved for the second eigenvalue in [18] ($P(\Omega)$ denotes the perimeter in the classical sense here):

Table 2.1 The optimal
Dirichlet eigenvalues λ_i^*, for
$i = 1, 2, \ldots, 15$ and the
multiplicity of the optimal
eigenvalue

i	Multiplicity	λ_i^*
1	1	18.17
2	2	36.34
3	3	46.13
4	3	64.30
5	2	78.15
6	3	88.48
7	3	106.12
8	3	118.88
9	3	132.34
10	4	142.69
11	4	159.40
12	4	172.88
13	4	186.91
14	4	198.94
15	5	209.62

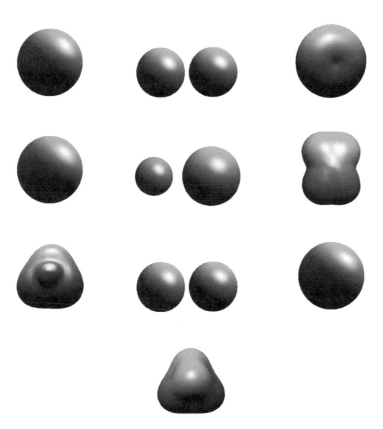

Fig. 2.10 The 3D minimizers of the first 10 Dirichlet eigenvalues (by courtesy of E. Oudet)

Theorem 2.66 (Bucur-Buttazzo-Henrot) *The problem*

$$\min\{\lambda_2(\Omega), \Omega \subset \mathbb{R}^2, P(\Omega) \leq P_0\}$$

has a solution Ω^*, this one is C^∞. Moreover the associated (normalized) eigenfunction u satisfies

$$|\nabla u|^2 = \frac{2\lambda_2}{P_0} \mathscr{C} \quad \text{on the boundary } \partial\Omega^* \tag{2.65}$$

where $\mathscr{C}$ is the curvature of the boundary. This implies that the curvature vanishes at two points and that $\partial\Omega$ does not contain neither segment, nor arc of circle.

Sketch of the Proof

- Existence of a minimizer follows easily using Theorems 2.24 and 2.28 and the fact that it suffices to look at convex domains.
- The C^∞ regularity can be proved using classical tools and a bootstrap argument. The proof consists in writing the boundary of the optimal domain as a graph of a function h, then writing the optimality condition we obtain an elliptic second order ordinary differential equation satisfied by h whose right-hand side involves the gradient of the eigenfunction. Using classical regularity results, both for this ODE and the PDE we get the result.
- To write the optimality condition (2.65), we use the domain derivative previously defined with a Lagrange multiplier to take into account the perimeter constraint. The precise value of the Lagrange multiplier can be obtained using the classical formulae (valid for any eigenvalue) $\lambda_2(\Omega) = \frac{1}{2} \int_{\partial\Omega} |\nabla u|^2 X.n \, ds$. A preliminary step consists in proving that the second eigenvalue is simple (in order to be differentiable with respect to the domain). For that purpose, we argue by contradiction: assuming that the eigenvalue is multiple (it can only be double here for a convex domain in the plane), we can obtain directional derivatives and we can always find, in that case, a deformation which makes λ_2 decrease.
- The qualitative properties are obtained thanks to this condition (2.65). The fact that the curvature has to vanish twice comes from the fact that the nodal line has to cross the boundary at two points for a convex domain, see [1, 46]. At last, the fact that the boundary cannot have a constant curvature is obtained exactly in the same way as (iii) of Theorem 2.33.

For the general case and in any dimension, namely for the problem

$$\min\{\lambda_k(\Omega), \Omega \subset \mathbb{R}^N, P(\Omega) \leq P_0\} \tag{2.66}$$

a recent paper gives existence and regularity, see [23]. Note that the expected regularity is weaker than in the two-dimensional case.

Theorem 2.67 (de Philippis-Velichkov) *The problem (2.66) has a solution. Its boundary is C^1 outside a closed set of Hausdorff dimension $N - 8$.*

Asymptotic Behaviour Let us denote by Ω_k^* an optimal domain for the problem (2.66) in the plane:

$$\lambda_k(\Omega_k^*) = \min\{\lambda_k(\Omega), \Omega \subset \mathbb{R}^2, P(\Omega) = 1\}.$$

A natural question is to wonder whether there exists a limit for the sequence Ω_k^* when $k \to \infty$. A positive answer is given in the paper [16]:

Theorem 2.68 (Bucur-Freitas 2015)   *The sequence of optimal planar domains* Ω_k^* *converges (for any classical distance) to the disk when* $k \to +\infty$.

Proof First of all, the sequence Ω_k^* being uniformly bounded, it has a limit for the Hausdorff distance. Let us denote by Ω_∞^* this limit. By continuity of the perimeter for the Hausdorff convergence among convex sets, $P(\Omega_\infty^*) = 1$. Let us introduced D the disk of perimeter 1. By definition of Hausdorff convergence, we have $\Omega_k^* \subset \Omega_\infty^* + B_\delta$, where B_δ is a disk of small radius δ. Thus

$$\lambda_k(\Omega_\infty^* + B_\delta) \le \lambda_k(\Omega_k^*) \le \lambda_k(D).$$

By Weyl's law:

$$\lambda_k(\Omega) \sim 4\pi k/|\Omega| \quad \text{when } k \to \infty.$$

This implies, for any $\delta > 0$

$$\frac{4\pi}{|\Omega_\infty^* + B_\delta|} \le \frac{4\pi}{|D|} \implies |D| \le |\Omega_\infty^*|.$$

We conclude thanks to the classical isoperimetric inequality.

Open Problem 2.69 *Prove a similar result in higher dimension. Prove a similar result with a volume constraint.*

2.4.3.2   Diameter Constraint

At last, let us consider the minimization problem with a diameter constraint in two-dimensions

$$\min\{\lambda_k(\Omega), \Omega \subset \mathbb{R}^2, D(\Omega) \le D_0\} \qquad (2.67)$$

where $D(\Omega)$ denotes the diameter. The same observation about the convex hull shows that the minimizer has to be convex, thus existence follows easily. Actually we can even prove that the minimizer is a convex body *of constant width*. In a recent work, see [7], with B. Bogosel and I. Lucardesi, we have considered this problem and, in particular, we identify the cases where the disk can be the minimizer or not.

The conclusion is a bit surprising. We give an exhaustive description of all the values of k for which the disk is (or is not) a local minimizer:

Theorem 2.70 *The eigenvalues for which the disk is a local minimizer for problem (2.67) are exactly*

$$\lambda_1, \lambda_2 = \lambda_3, \lambda_4 = \lambda_5, \lambda_7 = \lambda_8, \lambda_{11} = \lambda_{12},$$

$$\lambda_{16} = \lambda_{17}, \lambda_{27}, \lambda_{33} = \lambda_{34}, \lambda_{41} = \lambda_{42}, \lambda_{50}.$$

In all the other cases, the disk in not a local minimizer.

Acknowledgements This text is largely inspired by the book [29] of the author supplemented by more recent results. In that context, I am grateful to my collaborators Beniamin Bogosel, Ilaria Lucardesi, Davide Zucco who agree to include recent unpublished results.

References

1. G. Alessandrini, Nodal lines of eigenfunctions of the fixed membrane problem in general convex domains. Comment. Math. Helv. **69**(1), 142–154 (1994)
2. L. Ambrosio, A. Colesanti, E. Villa, Outer Minkowski content for some classes of closed sets. Math. Ann. **342**, 727–748 (2008)
3. P.R.S. Antunes, P. Freitas, Numerical optimization of low eigenvalues of the Dirichlet and Neumann Laplacians. J. Optim. Theory Appl. **154**(1), 235–257 (2012)
4. P.R.S. Antunes, P. Freitas, Optimal spectral rectangles and lattice ellipses. Proc. R. Soc. Lond. Ser. A Math. Phys. Eng. Sci. **469**(2150), 20120492 (2013)
5. C. Bandle, *Isoperimetric Inequalities and Applications*. Monographs and Studies in Mathematics, vol. 7 (Pitman, Boston, 1980)
6. A. Berger, The eigenvalues of the Laplacian with Dirichlet boundary condition in $\mathbb{R}^2$ are almost never minimized by disks. Ann. Global Anal. Geom. **47**(3), 285–304 (2015)
7. B. Bogosel, A. Henrot, I. Lucardesi, Minimization of $\lambda_k(\Omega)$ with a diameter constraint. SIAM J. Math. Anal. (2019, to appear)
8. L. Brasco, A. Pratelli, Sharp stability of some spectral inequalities. Geom. Funct. Anal. **22**(1), 107–135 (2012)
9. L. Brasco, G. De Philippis, B. Velichkov, Faber-Krahn inequalities in sharp quantitative form. Duke Math. J. **164**(9), 1777–1831 (2015)
10. H. Brézis, *Analyse Fonctionnelle* (Masson, Paris, 1983)
11. T. Briançon, Regularity of optimal shapes for the Dirichlet's energy with volume constraint. ESAIM: COCV **10**, 99–122 (2004)
12. T. Briançon, J. Lamboley, Regularity of the optimal shape for the first eigenvalue of the Laplacian with volume and inclusion constraints. Ann. Inst. H. Poincaré Anal. Non Linéaire **26**(4), 1149–1163 (2009)
13. D. Bucur, Regularity of optimal convex shapes. J. Convex Anal. **10**(2), 501–516 (2003)
14. D. Bucur, Minimization of the k-th eigenvalue of the Dirichlet Laplacian. Arch. Ration. Mech. Anal. **206**(3), 1073–1083 (2012)
15. D. Bucur, G. Buttazzo, *Variational Methods in Shape Optimization Problems*. Progress in Nonlinear Differential Equations and Their Applications (Birkhäuser, Boston, 2005)
16. D. Bucur, P. Freitas, Asymptotic behaviour of optimal spectral planar domains with fixed perimeter. J. Math. Phys. **54**(5), 053504 (2013)

17. D. Bucur, A. Henrot, Minimization of the third eigenvalue of the Dirichlet Laplacian. Proc. R. Soc. Lond. **456**, 985–996 (2000)
18. D. Bucur, G. Buttazzo, A. Henrot, Minimization of $\lambda_2(\Omega)$ with a perimeter constraint. Indiana Univ. Math. J. **58**, 2709–2728 (2009)
19. G. Buttazzo, G. Dal Maso, An existence result for a class of shape optimization problems. Arch. Rational Mech. Anal. **122**, 183–195 (1993)
20. T. Chatelain, M. Choulli, Clarke generalized gradient for eigenvalues. Commun. Appl. Anal. **1**(4), 443–454 (1997)
21. R. Courant, D. Hilbert, *Methods of Mathematical Physics*, vols. 1 and 2 (Wiley, New York, 1953 and 1962)
22. S.J. Cox, The generalized gradient at a multiple eigenvalue. J. Funct. Anal. **133**(1), 30–40 (1995)
23. G. De Philippis, B. Velichkov, Existence and regularity of minimizers for some spectral functionals with perimeter constraint. Appl. Math. Optim. **69**(2), 199C231 (2014)
24. Y. Egorov, V. Kondratiev, *On Spectral Theory of Elliptic Operators*. Operator Theory: Advances and Applications, vol. 89 (Birkhäuser, Basel, 1996)
25. G. Faber, Beweis, dass unter allen homogenen Membranen von gleicher Fläche und gleicher Spannung die kreisförmige den tiefsten Grundton gibt. Sitz. Ber. Bayer. Akad. Wiss. 169–172 (1923)
26. M. Flucher, Approximation of Dirichlet eigenvalues on domains with small holes. J. Math. Anal. Appl. **193**(1), 169–199 (1995)
27. G.H. Hardy, J.E. Littlewood, G. Pólya, *Inequalities, Reprint of the 1952 Edition, Cambridge Mathematical Library* (Cambridge University Press, Cambridge, 1988)
28. E.M. Harrell, P. Kröger, K. Kurata, On the placement of an obstacle or a well so as to optimize the fundamental eigenvalue. SIAM J. Math. Anal. **33**(1), 240–259 (2001)
29. A. Henrot, *Extremum Problems for Eigenvalues of Elliptic Operators*. Frontiers in Mathematics (Birkhäuser, Basel, 2006)
30. A. Henrot (ed.), *Shape Optimization and Spectral Theory* (De Gruyter Open, Warzaw, 2017). Downloadable at https://www.degruyter.com/view/product/490255
31. A. Henrot, E. Oudet, Minimizing the second eigenvalue of the Laplace operator with Dirichlet boundary conditions. Arch. Rational Mech. Anal. **169**, 73–87 (2003)
32. A. Henrot, M. Pierre, *Variation et optimisation de formes*. Mathématiques & Applications (Springer, Berlin, 2005). English version: *Shape Variation and Optimization*. Tracts in Mathematics, vol. 28 (European Mathematical Society, Zürich, 2018)
33. A. Henrot, D. Zucco, Optimizing the first Dirichlet eigenvalue of the Laplacian with an obstacle. https://arxiv.org/abs/1702.01307
34. J. Hersch, The method of interior parallels applied to polygonal or multiply connected membranes. Pacific J. Math. **13**, 1229–1238 (1963)
35. I. Hong, On an inequality concerning the eigenvalue problem of membrane. Kodai Math. Sem. Rep. **6**, 113–114 (1954)
36. B. Kawohl, *Rearrangements and Convexity of Level Sets in PDE*. Lecture Notes in Mathematics, vol. 1150 (Springer, Berlin, 1985)
37. S. Kesavan, On two functionals connected to the Laplacian in a class of doubly connected domains. Proc. R. Soc. Edinb. Sect. A **133**(3), 617–624 (2003)
38. S. Kesavan, *Symmetrization and Applications*. Series in Analysis, vol. 3 (World Scientific, Hackensack, 2006)
39. E. Krahn, Über eine von Rayleigh formulierte Minimaleigenschaft des Kreises. Math. Ann. **94**, 97–100 (1924)
40. E. Krahn, Über Minimaleigenschaften der Kugel in drei un mehr Dimensionen. Acta Comm. Univ. Dorpat. **A9**, 1–44 (1926)
41. J. Lamboley, About Hölder-regularity of the convex shape minimizing λ_2. Appl. Anal. **90**(2), 263–278 (2011)
42. J. Lamboley, A. Novruzi, Polygons as optimal shapes with convexity constraint. SIAM J. Control Optim. **48**, 3003–3025 (2009/2010)

43. J. Lamboley, A. Novruzi, M. Pierre, Regularity and singularities of optimal convex shapes in the plane. Arch. Ration. Mech. Anal. **205**, 311–343 (2012)
44. V. Maz'ja, S. Nazarov, B. Plamenevskii, *Asymptotische Theorie elliptischer Randwertaufgaben in singulär gestörten Gebieten. I.* Mathematische Lehrbücher und Monographien (Akademie-Verlag, Berlin, 1991)
45. D. Mazzoleni, A. Pratelli, Existence of minimizers for spectral problems. J. Math. Pures Appl. (9) **100**(3), 433–453 (2013)
46. A. Melas, On the nodal line of the second eigenfunction of the Laplacian in $\mathbb{R}^2$. J. Differ. Geom. **35**, 255–263 (1992)
47. S.A. Nazarov, J. Sokolowski, Asymptotic analysis of shape functionals. J. Math. Pures Appl. **82**(2), 125–196 (2003)
48. R. Osserman, The isoperimetric inequality. Bull. Am. Math. Soc. **84**(6), 1182–1238 (1978)
49. E. Oudet, Some numerical results about minimization problems involving eigenvalues. ESAIM COCV **10**, 315–335 (2004)
50. S. Ozawa, Singular variation of domains and eigenvalues of the Laplacian. Duke Math. J. **48**(4), 767–778 (1981)
51. G. Pólya, On the characteristic frequencies of a symmetric membrane. Math. Z. **63**, 331–337 (1955)
52. G. Pólya, G. Szegö, *Isoperimetric Inequalities in Mathematical Physics.* Annals of Mathematics Studies, vol. 27 (Princeton University Press, Princeton, 1951)
53. A.G. Ramm, P.N. Shivakumar, Inequalities for the minimal eigenvalue of the Laplacian in an annulus. Math. Inequal. Appl. **1**(4), 559–563 (1998)
54. B. Rousselet, Shape design sensitivity of a membrane. J. Optim. Theory Appl. **40**, 595–623 (1983)
55. V. Šverak, On optimal shape design. J. Math. Pures Appl. **72**(6), 537–551 (1993)
56. P. Tilli, D. Zucco, Asymptotics of the first Laplace eigenvalue with Dirichlet regions of prescribed length. SIAM J. Math. Anal. **45**, 3266–3282 (2013)
57. B.A. Troesch, Elliptical Membranes with smallest second eigenvalue. Math. Comput. **27**(124), 767–772 (1973)
58. S.A. Wolf, J.B. Keller, Range of the first two eigenvalues of the Laplacian. Proc. R. Soc. Lond. A **447**, 397–412 (1994)

Chapter 3
Topological Aspects of Critical Points and Level Sets in Elliptic PDEs

Alberto Enciso and Daniel Peralta-Salas

Abstract In this paper we discuss the emergence of topological structures through the critical points and level sets of solutions to elliptic PDEs. First, we analyze the local and global properties of the critical points of Green's functions on complete Riemannian manifolds, showing that the number of critical points on a surface of finite type admits a topological upper bound, a property that does not hold in higher dimensions. Second, we introduce two technical tools, Thom's isotopy theorem and a Runge-type global approximation theorem, which will allow us to construct solutions to a wide range of elliptic PDEs with level sets of complicated (sometimes bizarre) topologies. The model elliptic equation that we consider to illustrate the power of these techniques is the Helmholtz equation (monochromatic waves). These ideas are used in two seemingly unrelated applications: a 2001 conjecture of Sir Michael Berry about the existence of Schrödinger operators in Euclidean space with eigenfunctions having nodal lines of arbitrary knot type, and the construction of bounded solutions to the Allen-Cahn equation with level sets of any compact topology. The aim of these notes is not to provide detailed proofs of all the stated results but to introduce the main ideas and methods behind certain selected topics in the study of topological structures in PDEs. This is the set of lecture notes the second named author gave at the CIME Summer School held in Cetraro (Italy) during June 2017.

3.1 Introduction: Emergence of Topological Structures in Elliptic PDEs

The study of critical points and level sets of solutions to an elliptic PDE is a major topic in partial differential equations and geometric analysis, with connections that range from geometric measure theory to harmonic analysis and purely topological

A. Enciso · D. Peralta-Salas (✉)
Instituto de Ciencias Matemáticas, Consejo Superior de Investigaciones Científicas, Madrid, Spain
e-mail: aenciso@icmat.es; dperalta@icmat.es

© Springer Nature Switzerland AG 2018
C. Bianchini et al. (eds.), *Geometry of PDEs and Related Problems*,
Lecture Notes in Mathematics 2220, https://doi.org/10.1007/978-3-319-95186-7_3

aspects. Among the qualitative features of solutions to elliptic PDEs there are two aspects that have received considerable attention: symmetry in overdetermined boundary problems and convexity and starshapedness of the level sets of the solutions.

The origin of this kind of questions goes back to electrostatics and potential theory. There the key object is the electric potential, which is a scalar function satisfying the Poisson equation

$$\Delta u = f$$

in a domain $\Omega \subseteq \mathbb{R}^3$, where the function f describes the electric charge density. The level sets $u^{-1}(c)$ are called equipotential surfaces and the critical points $\{x \in \Omega : \nabla u(x) = 0\}$ are the equilibrium points of the electric field. If the domain Ω is not the whole space, boundary conditions need to be imposed (typically, in electrostatics one prescribes the value of the potential u on $\partial\Omega$ and suitable decay conditions at infinity). An analogous problem arises in the Newtonian theory of gravity, with u playing the role of the gravitational potential and the function f (which would now be nonpositive) describing the mass density.

It is remarkable that the study of the geometric and topological properties of the level sets and critical points of the electrostatic potential was actually pioneered by Faraday and Maxwell in the Nineteenth century. In this setting, the topology of the level sets of the electrostatic or gravitational potential is related to the "shape" of conducting or self-gravitating surfaces in equilibrium. The relevance for studying these objects of Faraday's lines of force, which are simply the integral curves of the gradient vector field $-\nabla u$, was unveiled in Maxwell's celebrated treatise [30], where they led to the first hints of what is nowadays known as Morse theory, which was introduced in order to study the number and location of critical (equilibrium) points of the field.

In this article we will be concerned with the number of critical points and the emergence of topological structures through the level sets of solutions to elliptic PDEs, focusing on solutions to linear equations or nonlinear equations in the linear regime, and on complex-valued eigenfunctions of Schrödinger-type operators. To present the questions that will be discussed in these notes, let us start with the most elementary context that one can consider: a harmonic function u in the Euclidean space $\mathbb{R}^n$:

$$\Delta u = 0.$$

A level set $u^{-1}(c)$ is regular if the gradient ∇u does not vanish at any point of $u^{-1}(c)$. It follows from the implicit function theorem that each connected component of a regular level set is an orientable codimension one C^∞ (actually C^ω) submanifold of $\mathbb{R}^n$ without boundary. Moreover, an easy argument using the maximum principle implies that all these connected components must be noncompact. Apart from these properties, until very recently there were no results on the admissible topological types of the level sets of a harmonic function. For

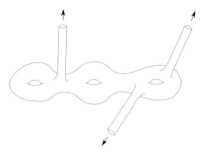

Fig. 3.1 A torus of genus $g = 3$ and $N = 3$ ends

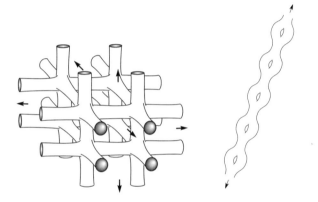

Fig. 3.2 From the left to the right, an infinite jungle gym and a torus of infinite genus

instance, in dimension $n = 3$ it is easy to construct explicit harmonic functions having level sets with a connected component diffeomorphic to elementary surfaces such as the plane or the cylinder, but it is an entirely different matter to ascertain whether there can be more complicated surfaces, as in the following:

Question 3.1 Is there a harmonic function u in $\mathbb{R}^3$ such that $u^{-1}(0)$ has a connected component diffeomorphic to the genus g torus with N ends? (cf. Fig. 3.1).

Question 3.2 Is there a harmonic function u in $\mathbb{R}^3$ such that $u^{-1}(0)$ has a connected component diffeomorphic to the torus of infinite genus or to the infinite jungle gym? (cf. Fig. 3.2).

One can also consider the transverse intersection of the zero sets of $m \geqslant 2$ harmonic functions. Unlike a single level set, these intersections can be compact. Here one would be interested in sets diffeomorphic to pathological objects as in the following

Question 3.3 Let Σ be an exotic sphere of dimension 7, so that Σ is a smooth manifold homeomorphic to the standard 7-sphere $\mathbb{S}^7$ but not diffeomorphic to it. It is known that Σ can be smoothly embedded in $\mathbb{R}^{14}$. Are there seven harmonic functions $u_1, \ldots, u_7$ on $\mathbb{R}^{14}$ such that the exotic sphere Σ is diffeomorphic to a component of the joint level set $u_1^{-1}(0) \cap \cdots \cap u_7^{-1}(0)$?

These questions illustrate the main problem that we shall address in these notes: do there exist solutions to an elliptic PDE with level sets of prescribed topology?. The theory dealing with this kind of questions was first introduced by the authors in [13] and [15] to study, respectively, the vortex lines of stationary solutions of inviscid, incompressible fluids, and the level sets of harmonic functions. In order to avoid the technical difficulties that are present in the study of the level sets of harmonic functions due to their noncompactness, in this paper we shall focus on the study of compact level sets of solutions to the Helmholtz equation

$$\Delta u + u = 0$$

in $\mathbb{R}^n$, and related equations. For completeness, we will mention that, surprisingly enough, the answer to all questions stated above regarding harmonic functions turns out to be positive [15].

This article is organized as follows. In Sect. 3.2 we will start by studying the critical points of Green's functions on complete manifolds, where we will prove a topological upper bound for the number of critical points on surfaces, and that this property does not hold in higher dimensions. The next sections will be devoted to the topology of level sets. In Sect. 3.3 we will present the general strategy introduced in [13] and [15] to approach this kind of problems, and we shall introduce two technical tools that are key for this strategy, Thom's isotopy theorem and a Runge-type global approximation theorem. In Sect. 3.4, these techniques will be applied to construct monochromatic waves (i.e. solutions to Helmholtz equation) in $\mathbb{R}^n$ with prescribed compact level sets. Since the Helmholtz equation models, as a rescaled limit, a wide range of phenomena in spectral theory and nonlinear PDEs, we will present two non trivial applications of the previous results. First, in Sect. 3.5, we will construct complex-valued high-energy eigenfunctions of Schrödinger operators in $\mathbb{R}^3$ with nodal sets of arbitrary knot type. Second, in Sect. 3.6 we will consider level sets for nonlinear equations close to the linear regime, illustrating the ideas with the Allen–Cahn equation.

3.2 Critical Points of Green's Functions on Complete Manifolds

The study of the topology of the level sets of a function u is intimately connected with the study of its critical points, that is, with the locus where its gradient ∇u vanishes. Indeed, it is well known that the critical values of u (i.e., the values that u takes at its critical points) are those at which the compact regular level sets $u^{-1}(c)$ can change its topology. Essentially, the reason for this is that if the level set $u^{-1}(c)$ is compact for all $c_1 \leqslant c \leqslant c_2$ and u does not have any critical values in the interval

$[c_1, c_2]$, the vector field $X := \nabla u / |\nabla u|^2$ is well defined in this set and satisfies

$$X \cdot \nabla u = 1 \,,$$

so its time t flow maps $u^{-1}(c_1)$ onto $u^{-1}(c_1 + t)$ for all $0 < t < c_2 - c_1$.

To illustrate the study of the critical points of solutions to an elliptic PDE, we will consider the case of the Green's function G (with pole y) of the Laplace-Beltrami operator of a noncompact n-dimensional manifold M endowed with a Riemannian metric g. An important difficulty in the study of critical points lies in the fact that Green's function estimates are not sufficiently fine to elucidate whether the gradient of G vanishes in a certain region. Moreover, it is well known that the codimension of the critical set of G is at least 2 [26], which introduces additional complications in the analysis. For these reasons, the results on critical points of Green's functions are surprisingly scarce. All along this section $\mathrm{Cr}(f)$ denotes the critical set of a function f, and we shall denote by Δ_g, ∇_g, $|\cdot|_g$ and dist_g, respectively, the Laplacian, gradient operator, norm and distance function in (M, g),

3.2.1 Li-Tam Green's Functions

As is usual in the literature, we will assume that the Green's function satisfies certain convenient conditions at infinity in order to control some global properties of its level sets; in particular, a standard choice is to restrict one's attention to the minimal Green's function whenever it exists. More generally, we shall assume in what follows that the Green's function has been obtained through an exhaustion procedure as in Li and Tam's paper [29].

Let us summarize the main facts about Li–Tam Green's functions. We will henceforth denote by $\mathscr{G}_\Omega : (\Omega \times \Omega) \setminus \mathrm{diag}(\Omega \times \Omega) \to \mathbb{R}$ the symmetric Dirichlet Green's function of a bounded domain $\Omega \subset M$, which is defined by

$$\Delta_g \mathscr{G}_\Omega(\cdot, y) = -\delta_y \quad \text{in } \Omega \,, \qquad \mathscr{G}_\Omega(\cdot, y) = 0 \quad \text{on } \partial\Omega \,, \tag{3.1}$$

and set $G_\Omega := \mathscr{G}_\Omega(\cdot, y)$. Denoting by $\Omega_1 \subset \Omega_2 \subset \cdots$ an exhaustion of M by bounded domains, it was proved in [29] that there exists a sequence of nonnegative real numbers $(a_j)_{j=1}^{\infty}$ such that $G_{\Omega_j} - a_j$ converges uniformly on compact sets of $M \setminus \{y\}$ to a Green's function G with pole y, and that it coincides with the minimal one whenever the latter exists. We will refer to the class of Green's functions that arise through this procedure as *Li–Tam Green's functions*. Li–Tam Green's functions are generally non-unique, but in any case a Li–Tam Green's function has the following properties:

(i) G is decreasing in the sense that

$$\sup_{M \setminus B_g(y,r)} G = \max_{\partial B_g(y,r)} G$$

for all $r > 0$, $B_g(y, r)$ being the geodesic ball centered at the pole y of radius r.
(ii) $\mathcal{G}$ is a symmetric function, that is, $\mathcal{G}(x, y) = \mathcal{G}(y, x)$.

3.2.2 A Topological Upper Bound on Surfaces

In [14] we proved that any Li-Tam Green's function of a noncompact surface M of finite type has a finite number of critical points depending only on the topology of the surface. We recall that a surface is of *finite type* if its first Betti number $b_1(M)$ is finite. These surfaces can be described with two numbers: their genus v and their number of ends λ, in which case $b_1(M) = 2v + \lambda - 1$.

Theorem 3.4 *Let (M, g) be a smooth noncompact Riemannian surface of finite type. The number of critical points of any Li–Tam Green's function G on M is bounded as*

$$|Cr(G)| \leqslant b_1(M) = 2v + \lambda - 1,$$

and if this upper bound is attained then G is Morse (i.e. all its critical points are nondegenerate).

The proof of this theorem is based on an index-theoretical argument that combines a local analysis and a global one. The local part consists of the calculation of the Hopf index of a critical point of the Green's function using a blow-up technique. The global part exploits the conformal symmetry of the equation to show that the critical set of G is finite and analyze the asymptotic behavior of the Green's function in each end of the surface.

Let us begin with the following local lemma, where we estimate the index that a critical point of G can have. We recall that the index of a critical point is the degree of a certain map defined by the gradient of G. This result is contained in [2], and is analogous to Cheng's local analysis of the nodal set of the eigenfunctions on a surface [7], however the proof we present below relies on arguments that are typical in the study of the integral curves of a vector field (blow-up, hyperbolicity and Bendixson's theorem).

Lemma 3.5 *Let $z \in M$ be a critical point of the Green's function G, and call $m \geqslant 2$ the degree of the lowest nonzero homogeneous term in the Taylor expansion of $G - G(z)$ at z. Then z is an isolated zero of $\nabla_g G$ and its index is $\mathrm{ind}(z) = 1 - m$.*

Proof The result being local, we will prove it using local isothermal coordinates $x = (x_1, x_2)$. In this coordinate chart we can safely assume that $z = 0$ and the metric tensor reads as a smooth matrix-valued function g_{ij} of the form $f(x_1, x_2)\delta_{ij}$, with f a positive smooth function. With some abuse of notation, we will still denote by G the expression of the Green's function in the coordinate system. The conformal invariance of the Laplacian implies that G is harmonic with respect to the

Euclidean metric in the coordinates (x_1, x_2), i.e. it satisfies the equation $\Delta G = 0$ in a neighborhood of 0, where Δ is the Euclidean Laplacian.

Since G is an analytic (C^ω) function in these coordinate, $G - G(0)$ can vanish only up to finite order at 0, so m is necessarily finite. Then there exists a homogeneous polynomial h_m of degree m such that

$$G(x) - G(0) = h_m(x) + O(|x|^{m+1}), \tag{3.2a}$$

$$\nabla_g G(x) = f^{-1} \nabla G = f^{-1}(\nabla h_m(x) + O(|x|^m)). \tag{3.2b}$$

It is straightforward to check that the polynomial h_m is harmonic with respect to the Euclidean metric ($\Delta h_m = 0$), which implies that the origin is an isolated critical point of h_m.

Let us now consider polar coordinates $(r, \theta) \in \mathbb{R}^+ \times \mathbb{S}^1$ defined by $(x_1, x_2) = (r \cos \theta, r \sin \theta)$. As h_m is harmonic, it readily follows that in these coordinates one has $h_m(r, \theta) = C_0 r^m \cos(m\theta - \theta_0)$ for some constants C_0 and θ_0. There is obviously no loss of generality in setting $\theta_0 = 0$. We define the polar blow up of the gradient $\nabla_g G$ at 0 using polar coordinates as the vector field

$$X := \frac{f}{C_0 m \, r^{m-2}} \nabla_g G = \frac{1}{C_0 m \, r^{m-2}} (\nabla h_m + O(r^m)),$$

where we have used Eq. (3.2b). The blown-up integral curves of $\nabla_g G$ are then given by

$$\dot{r} = r \cos m\theta + O(r^2), \tag{3.3a}$$

$$\dot{\theta} = -\sin m\theta + O(r). \tag{3.3b}$$

The blown-up critical points are thus $(0, \theta_k)$, with $\theta_k := k\pi/m$ and $k = 1, \ldots, 2m$. The Jacobian matrix of X at $(0, \theta_k)$ is

$$DX(0, \theta_k) = \begin{pmatrix} (-1)^k & 0 \\ 0 & (-1)^{k+1} \end{pmatrix}, \tag{3.4}$$

so these critical points are hyperbolic saddles. By blowing down, we immediately find that a deleted neighborhood of 0 consists exactly of $2m$ hyperbolic sectors of the vector field X.

Since the field X is proportional to the gradient field $\nabla_g G$ through a nonvanishing factor, we conclude that the critical point of G is isolated, and we can compute its index using Bendixson's formula for the index of a planar vector field [31]:

$$\text{ind}(0) = 1 - \frac{\text{number of hyperbolic sectors}}{2} = 1 - m,$$

as claimed.

Let us now focus on global aspects of the problem. Since (M, g) is a smooth Riemannian surface of finite type, the uniformization theorem implies that there is a compact surface Σ endowed with a metric of constant curvature $\bar{g}$, a certain number $\lambda_1 \geqslant 0$ of isolated points $p_i \in \Sigma$ and another number $\lambda_2 \geqslant 0$ of topological disks $D_i \subset \Sigma$ such that (M, g) is conformally isometric to the interior of

$$\overline{M} := \Sigma \backslash \left(\bigcup_{i=1}^{\lambda_1} \{p_i\} \cup \bigcup_{j=1}^{\lambda_2} D_j \right)$$

with the metric $\bar{g}$. Equivalently, there exists a diffeomorphism $\Phi : M \to \text{int } \overline{M}$ and a smooth positive function f on M such that $\Phi^* \bar{g} = fg$. Moreover, the boundaries ∂D_i can be taken analytic without loss of generality. Denoting the genus of Σ by v and setting $\lambda := \lambda_1 + \lambda_2 \geqslant 1$, it is clear that the pair (v, λ) determines the surface M uniquely up to diffeomorphism.

In the following lemma we prove several basic properties of the Li–Tam Green's function G. In terms of the diffeomorphism Φ considered above, it is convenient to introduce the notation $\overline{G} := G \circ \Phi^{-1}$ for the Green's function on (the interior of) $\overline{M}$ and use the notation $\bar{y} := \Phi(y)$ for the image of the pole of the Green's function under the aforementioned diffeomorphism. It should be noted too that obviously $\text{Cr}(G) = \Phi^{-1}(\text{Cr}(\overline{G}))$.

Lemma 3.6 *The Green's function has the following properties:*

(i) *G is positive if and only if $\lambda_2 \geqslant 1$. In this case, $\overline{G}$ tends to zero at the boundary of each disk ∂D_j. $\overline{G}$ can be extended so as to satisfy $\Delta_{\bar{g}} \overline{G} = 0$ in a neighborhood of each isolated point p_i and each circle ∂D_j, and the gradient $\nabla_{\bar{g}} \overline{G}$ is orthogonal to (and nonzero at) ∂D_j.*

(ii) *If $\lambda_2 = 0$, $\overline{G}$ tends to $-\infty$ at, at least, one of the points p_i, and satisfies the equation $\Delta_{\bar{g}} \overline{G} = 0$ in a neighborhood of any point p_j where $\overline{G}$ is bounded from below.*

(iii) *$\overline{G}$ has a finite number of critical points in the interior of $\overline{M} \backslash \{\bar{y}\}$.*

Proof A short calculation using the conformal properties of the Laplacian in two dimensions shows that $\overline{G}$ satisfies the equation

$$\Delta_{\bar{g}} \overline{G} = -\delta_{\bar{y}}$$

in the interior of $\overline{M}$. When the number λ_2 of disks is at least 1, it is standard (for example, due to the existence of bounded positive harmonic functions in the interior of $\overline{M}$ [29]) that G is the minimal Green's function of (M, g), which corresponds to the unique solution $\overline{G}$ of the boundary problem

$$\Delta_{\bar{g}} \overline{G} = -\delta_{\bar{y}} \quad \text{in } \Sigma \backslash \bigcup_{j=1}^{\lambda_2} \overline{D_j}, \qquad \overline{G} = 0 \quad \text{on } \partial D_j \text{ for all } j.$$

Moreover, the fact that $\bar{g}$ and ∂D_j are analytic ensures that $\overline{G}$ is analytic and harmonic in a neighborhood of each circle ∂D_j.

When $\lambda_2 = 0$, the function $\overline{G}$ satisfies the equation $\Delta_{\bar{g}} \overline{G} = 0$ but at the pole $\bar{y}$ and the isolated points p_i. By property (i) of Li–Tam Green's functions, $\overline{G}$ is upper bounded at each point p_i. If it is also lower bounded, it is standard that p_i is a removable singularity and $\Delta_{\bar{g}} \overline{G} = 0$ in a neighborhood of p_i. If $\overline{G}$ is not lower bounded, p_i is an isolated singularity of $\overline{G}$, and the fact that $\overline{G}$ is upper bounded readily implies that $\Delta_{\bar{g}} \overline{G} = c_i \delta_{p_i}$ in a neighborhood of p_i for some positive constant c_i. Hence

$$\Delta_{\bar{g}} \overline{G} = -\delta_{\bar{y}} + \sum_{i=1}^{\lambda_1} c_i \, \delta_{p_i}$$

in the closed manifold Σ, so obviously $\sum_i c_i = 1$.

Let us now prove that the number of critical points of G is finite. As the boundary ∂D_j is analytic and $\overline{G}$ is positive (when $\lambda_2 > 0$), by Hopf's boundary point lemma there is a neighborhood U_j of each disk $\overline{D_j}$ such that $\overline{G}$ does not have any critical points in U_j. The boundary ∂D_j being a level curve of $\overline{G}$, this ensures the gradient $\nabla_{\bar{g}} \overline{G}$ is orthogonal to this set. If p denotes either an isolated point p_i where $\overline{G}$ tends to $-\infty$ or the pole $\bar{y}$, the well known asymptotics

$$\left| G(x) + C_p \log \mathrm{dist}_{\bar{g}}(x, p) \right| + \left| \nabla_{\bar{g}} \overline{G}(x) + C_p \nabla_{\bar{g}} \log \mathrm{dist}_{\bar{g}}(x, p) \right|_{\bar{g}} = O(1), \quad (3.5)$$

for the Green's function ensures that $\overline{G}$ does not have any critical points in a neighborhood of p. Here C_p is to be interpreted as $-c_i/2\pi$ if $p = p_i$ and $1/2\pi$ if $p = \bar{y}$. Since the critical points of $\overline{G}$ are isolated and cannot accumulate at the isolated points p_i where $\overline{G}$ is bounded due to the fact that $\overline{G}$ is harmonic at these points and Lemma 3.5, it follows that $\overline{G}$ has a finite number of critical points, as claimed.

We are now ready to prove Theorem 3.4. In the proof we will assume that the Green's function $\overline{G}$ has been extended to the points p_i where it is bounded and to the circles ∂D_j as in Lemma 3.6.

Proof (Proof of Theorem 3.4) Let $D_{\bar{y}}$ be a small disk in $\overline{M}$ that contains the pole $\bar{y}$. For concreteness, using the asymptotics for the Green's function (3.5) we can assume that the boundary of this disk is the level curve $\overline{G}^{-1}(c_0)$ for some large positive constant c_0. By construction, the gradient $\nabla_{\bar{g}} \overline{G}$ is transverse (in fact, orthogonal) to $\partial D_{\bar{y}}$. By the asymptotics for the gradient (3.5) it is clear that $\overline{G}$ does not have any critical points in the closure of $D_{\bar{y}}$.

Suppose that the number λ_2 of removed disks is at least one and consider the manifold with boundary

$$\Sigma' := \Sigma \setminus \left(D_{\bar{y}} \cup \bigcup_{j=1}^{\lambda_2} D_j \right).$$

By item (i) in Lemma 3.6 the vector field $\nabla_{\bar{g}}\overline{G}$ is smooth in Σ' and transverse to its boundary. Let us denote by $z_1, \ldots, z_N$ the critical points of $\overline{G}$ in $\overline{M}$, which are finite in number by item (iii) in Lemma 3.6. In addition to this, some of the isolated points p_i can be critical points of $\overline{G}$ too; without loss of generality we can order them so that these critical points are $p_1, \ldots, p_{\lambda_1'}$ with $0 \leqslant \lambda_1' \leqslant \lambda_1$.

If we now apply Hopf's index theorem to the vector field $\nabla_{\bar{g}}\overline{G}$ in the manifold with boundary Σ' and realize that in dimension 2 it does not matter whether the vector field points inward or outward at the boundary, we get that the sum of the indices of the zeros of $\nabla_{\bar{g}}\overline{G}$ equals the Euler characteristic of Σ':

$$\sum_{k=1}^{N} \mathrm{ind}(z_k) + \sum_{i=1}^{\lambda_1'} \mathrm{ind}(p_i) = \chi(\Sigma').$$

Since $\chi(\Sigma')$ can be readily shown to be $1 - 2\nu - \lambda_2$ and the index of each critical point is smaller than or equal to -1 by Lemma 3.5, we find

$$-\sum_{k=1}^{N} \mathrm{ind}(z_k) = 2\nu + \lambda_2 - 1 + \sum_{i=1}^{\lambda_1'} \mathrm{ind}(p_i) \leqslant 2\nu + \lambda_2 - \lambda_1' - 1 \leqslant 2\nu + \lambda - 1 . \quad (3.6)$$

This implies that $N \leqslant 2\nu + \lambda - 1$, and the equality is not satisfied but perhaps when $\mathrm{ind}(z_k) = -1$ for all k, that is, when $\overline{G}$ is Morse. (The fact that the critical point z_k is nondegenerate when $\mathrm{ind}(z_k) = -1$ is an immediate consequence of Lemma 3.5.)

Consider now the case where $\lambda_2 = 0$. We have seen in item (ii) of Lemma 3.6 that there are some points p_j where $\overline{G}$ tends to $-\infty$, and that $\overline{G}$ is harmonic at the other points p_k. Among the points where the function is harmonic, some can be critical points of $\overline{G}$. Without loss of generality we can label these critical points as $p_1, \ldots, p_{\lambda_1'}$. Let λ_1'' be the number of points where $\overline{G}$ tends to $-\infty$; for concreteness, we can assume that these points are $p_{\lambda_1'+1}, \ldots, p_{\lambda_1'+\lambda_1''}$. Notice that necessarily $\lambda_1'' \geqslant 1$.

As we did with the pole $\bar{y}$, let us take small disks D_{p_j} containing the points p_j, with $\lambda_1' + 1 \leqslant j \leqslant \lambda_1' + \lambda_1''$. As before, the boundary circles can be chosen as the λ_1'' components of the level curve $\overline{G}^{-1}(-c_0)$ for large enough c_0. The asymptotics for the Green's function given by (3.5) ensures that $\overline{G}$ does not have any critical points in the closure of these disks and $\nabla_{\bar{g}}\overline{G}$ is transverse (and orthogonal by construction) to the boundary ∂D_{p_j}.

We can now apply Hopf's index theorem to the vector field $\nabla_{\bar{g}}\overline{G}$ in the manifold with boundary

$$\Sigma'' := \Sigma \setminus \left(D_{\bar{y}} \cup \bigcup_{j=\lambda_1'+1}^{\lambda_1'+\lambda_1''} D_{p_j} \right).$$

The Euler characteristic of Σ'' is $1 - 2\nu - \lambda_1''$, so denoting the critical points of $\overline{G}$ in $\overline{M}$ again by $z_1, \ldots, z_N$ and arguing as above one arrives at the inequality

$$-\sum_{k=1}^{N} \text{ind}(z_k) = 2\nu + \lambda_1'' - 1 + \sum_{i=1}^{\lambda_1'} \text{ind}(p_i) \leqslant 2\nu + \lambda_1'' - \lambda_1' - 1 \leqslant 2\nu + \lambda - 1.$$

This yields $N \leqslant 2\nu + \lambda - 1$, the inequality being saturated at most when $\text{ind}(z_k) = -1$ for all k (that is, when $\overline{G}$ is Morse).

Remark 3.7 In addition to the proof of Theorem 3.4 presented above following [14], a different, more visual proof of this result that hinges on the qualitative properties of the flow defined by the gradient of G can be found in [18]. In this second approach, the result follows from a more general heuristic principle, which informally asserts that M can be decomposed as the union of a topological disk D and a (possibly disconnected) noncompact graph $\mathscr{F}$, both of which consist of integral curves of ∇G. The disk is the union of the pole y and the points of M whose ω-limit along the flow of ∇G is y. We recall that the ω-limit of a point $x_0 \in M$ along the flow $\varphi_t(x_0)$ of a vector field is the set $\{x \in M : \varphi_{t_k}(x_0) \to x$ for some sequence $t_k \to \infty\}$. The graph $\mathscr{F}$ consists of the critical points of G, their stable components, and certain integral curves of ∇G that escape to infinity. When suitably compactified, $\mathscr{F}$ is a connected graph that encodes the topology of the surface, the rank of the first homology group of $\mathscr{F}$ being twice the genus of M. The bounds for the number of critical points of the Green's function stem from the structure of the graph $\mathscr{F}$.

3.2.3   Critical Points in Higher Dimensions

It is remarkable that the picture is completely different in higher dimensions. Indeed, in [14] we proved that on any n-dimensional noncompact manifold M, $n \geqslant 3$, there exists a Riemannian metric g whose Green's function has an arbitrarily high number of critical points. It is worth emphasizing that the way we manage to control the critical points is precisely by proving the existence of level sets of G of complicated topology and showing that, in a precise sense, one can generically assume that the critical points of the Green's function are nondegenerate in a certain region of space. For simplicity, let us state this result in the case where M is diffeomorphic to $\mathbb{R}^n$:

Theorem 3.8 *Given any integer N, there is a complete Riemannian metric g on $\mathbb{R}^n$, with $n \geqslant 3$, whose minimal Green's function has at least N nondegenerate critical points.*

The key step of the proof of this result is to show that one can choose a metric on $\mathbb{R}^n$ whose minimal Green's function approximates the (Euclidean) Dirichlet Green's function of any fixed bounded domain. More precisely, given a bounded domain Ω in $\mathbb{R}^n$ containing the point y, let us denote by G_Ω its Dirichlet Green's function with a pole at the point y, which satisfies the equation

$$\Delta G_\Omega = -\delta_y \quad \text{in } \Omega , \qquad G_\Omega|_{\partial\Omega} = 0$$

with Δ the ordinary Laplacian of $\mathbb{R}^n$.

Lemma 3.9 *Let Ω be a bounded domain of $\mathbb{R}^n$ with connected boundary. Given a compact subset S of $\Omega\backslash\{y\}$, for any $\delta > 0$ there is a complete conformally flat metric g on $\mathbb{R}^n$ whose minimal Green's function G tends to zero at infinity and satisfies*

$$\|G - G_\Omega\|_{C^2(S)} < \delta .$$

To see that Lemma 3.9 leads to Theorem 3.8, one starts by proving that, by perturbing the domain Ω a little if necessary, one can assume that the Green's function G_Ω is Morse (that is, all its critical points are nondegenerate). This is not hard to believe but the proof is not immediate. Since the gradient of G_Ω points inwards (and in particular it does not vanish) on the boundary of Ω by Hopf's boundary point lemma, one can apply Morse theory for manifolds with boundary to the auxiliary C^2 function $f : \overline{\Omega} \to \mathbb{R}$ defined by

$$f(x) := \begin{cases} -1/(G_\Omega(x) + 1)^2 & \text{if } x \neq y , \\ 0 & \text{if } x = y . \end{cases}$$

It then follows that f has at least $b_p(\overline{\Omega})$ critical points of index $n - p$, for all $0 \leqslant p \leqslant n$, where b_p denotes the pth Betti number. It is a straightforward computation that all the critical points of f other than y are also critical points of G_Ω and they have the same Morse indices. Since any C^2-small perturbation of a Morse function is Morse and has the same number of critical points, to prove Theorem 3.8 it is then enough to apply the statement of Lemma 3.9, for instance, to a domain Ω in $\mathbb{R}^n$ whose boundary is the connected sum of N copies of $\mathbb{S}^1 \times \mathbb{S}^{n-2}$, for in this case its first Betti number is $b_1(\overline{\Omega}) = N$.

To prove Lemma 3.9 we take, for any positive integer j, a smooth function $\varphi_j : \mathbb{R}^n \to [1, \infty)$ such that $\varphi_j(x) = 1$ if $x \in \Omega$ and $\varphi_j(x) = j$ if $\text{dist}(x, \Omega) > \frac{1}{j}$. Let us now define the conformally flat metrics $g_j := \varphi_j g_0$ on $\mathbb{R}^n$, where g_0 denotes the Euclidean metric. As the metric is flat outside a compact set, it can be proved that there is a unique minimal Green's function G_j with a pole at a point y. For simplicity of notation, let us choose the origin of coordinates so that $y = 0$. The idea now is that, as the metric is very large outside Ω, the Green's function tends to zero outside Ω, so it tends to G_Ω as $j \to \infty$. The way we make this precise is via the variational formulation of the problem, the interested reader can consult a detailed proof in [14].

We finish this section with the following corollary, which is a straightforward consequence of Lemma 3.9. It shows that the level sets of a minimal Green's function on a conformally flat Riemannian manifold of dimension $n \geqslant 3$ can exhibit any compact topology.

Corollary 3.10 *Let* $\Sigma \subset \mathbb{R}^n$ *be a compact, codimension one submanifold without boundary of class* C^∞, *and let* y *be a point in the domain bounded by* Σ. *Then, there exists a complete conformally flat Riemannian manifold* $(\mathbb{R}^n, g)$ *whose minimal Green's function with pole* y *tends to zero at infinity and has a level set diffeomorphic to* Σ. *The diffeomorphism is as close to the identity as one wishes.*

Remark 3.11 An analogous theorem holds in the context of low-energy eigenfunctions of the Laplace-Beltrami operator on a compact Riemannian manifold M of dimension $n \geqslant 3$. More precisely, in [17] we proved that, for any compact, codimension one submanifold $\Sigma \subset M$ separating and without boundary, there exists a metric g on M such that its first nontrivial eigenvalue is simple and the nodal set of the corresponding eigenfunction is diffeomorphic to Σ. We say that a submanifold Σ is separating if its complement $M \backslash \Sigma$ is the union of two disjoint open sets. This result is valid even for metrics within a conformal class and with fixed volume [20].

3.3 General Strategy and Two Technical Tools: Thom's Isotopy Theorem and a Runge-Type Global Approximation Theorem

In Sect. 3.2, Corollary 3.10, we constructed metrics on any noncompact manifold of dimension $n \geqslant 3$ whose minimal Green's functions have a level set of prescribed topology. A change in the metric is a change in the coefficients of the Laplace-Beltrami operator, so we play with a lot of freedom: we are changing the equation.

The questions stated in Sect. 3.1 are of a totally different nature. For concreteness, in this section we shall focus on the Helmholtz equation $\Delta u + u = 0$ in $\mathbb{R}^n$. The equation (and in particular the metric) is fixed and the freedom is the space of solutions to the equation (which is infinite dimensional). The problem is then how flexible/rigid the solutions to the Helmholtz equation are regarding the topology of their level sets:

Question 3.12 Let Σ be a codimension one compact submanifold of $\mathbb{R}^n$ (without boundary and orientable). Is there a solution u to the Helmholtz equation $\Delta u + u = 0$ in $\mathbb{R}^n$ and a diffeomorphism $\Phi : \mathbb{R}^n \rightarrow \mathbb{R}^n$ such that $\Phi(\Sigma)$ is a connected component of the nodal set $u^{-1}(0)$?.

The key point in this question is that u is a solution to the PDE in the whole space. Indeed, for any analytic hypersurface Σ of $\mathbb{R}^n$, one has $\Sigma = v^{-1}(0)$ for some function satisfying $\Delta v + v = 0$ in a neighborhood of Σ (it is enough to take v as the local solution, given by the Cauchy–Kowalewski theorem, to the problem $\Delta v + v = 0$, $v|_\Sigma = 0$, $\partial_\nu v|_\Sigma = 1$), but this function v cannot be generally extended as a global solution to the Helmholtz equation.

Remark 3.13 It is essential to allow the level sets to be realized only modulo diffeomorphism: for example, it is known [24] that the curves $x_2 = x_1^s$, which are all diffeomorphic for any integer $s \geqslant 1$, can be a connected component of the zero level set of a harmonic function in $\mathbb{R}^2$ if and only if $s = 1$ or $s = 2$.

In [13] and [15] we introduced a new strategy to address this problem (and analogous questions for other, not necessarily elliptic, PDEs). The basic strategy to construct a solution to a PDE such that a hypersurface Σ is diffeomorphic to a connected component of one of its level sets is the following. We first construct a function v that satisfies the equation in a neighborhood of Σ and has a level set diffeomorphic to Σ. This can be done e.g. using the Cauchy-Kowalewski theorem or an appropriately chosen boundary value problem. Second, we will approximate this local solution v by a global solution u using a Runge-type global approximation theorem. Hence, the key for the success of this strategy is to be able to ensure that the level set of the local solution is "robust" in the sense that it is preserved, up to a diffeomorphism, under the perturbation corresponding to the above global approximation. This structural stability can be accomplished using Thom's celebrated isotopy theorem if the level sets are compact. In view of this, the local solution v must be constructed in such a way that the above stability condition holds.

3.3.1 Thom's Isotopy Theorem

When the level sets of a function are compact, their structural stability under small perturbations is granted by Thom's isotopy theorem [1, 33]. This result provides a sufficient condition in order that a connected component Σ of the level set $v^{-1}(c)$ of a smooth function v be preserved, up to a diffeomorphism, by C^k-small perturbations of v. The condition is that the gradient ∇v does not vanish at any point of Σ. In general, if $v = (v_1, \cdots, v_m)$ is a smooth map $\mathbb{R}^n \to \mathbb{R}^m$, with $n > m$, Thom's isotopy theorem reads as follows:

Theorem 3.14 *Let $v : \mathbb{R}^n \to \mathbb{R}^m$ be a smooth map, and consider a compact connected component Σ of the level set $v^{-1}(c)$. Suppose that*

$$rank\,(\nabla v_1, \cdots, \nabla v_m)|_\Sigma = m \,. \tag{3.7}$$

Then, for any $\epsilon > 0$, there exists some $\delta > 0$ such that for any smooth map $u :$ $\mathbb{R}^n \to \mathbb{R}^m$ with $\|u - v\|_{C^k(U)} < \delta$ ($k \geqslant 1$), where U is a neighborhood of Σ, one can transform Σ by a diffeomorphism Φ of $\mathbb{R}^n$ so that $\Phi(\Sigma)$ is the intersection of the level set $u^{-1}(c)$ with U. Moreover, the diffeomorphism Φ only differs from the identity in a proper subset of U and satisfies $\|\Phi - id\|_{C^k(\mathbb{R}^n)} < \epsilon$.

This theorem establishes the structural stability of the compact (connected components of the) level sets of a smooth function v that are regular in the sense of condition (3.7). In light of the strategy presented above to prescribe the topology of the level sets of a solution to the Helmholtz equation, this result is useful not only to preserve the topology when applying the Runge-type global approximation theorem (which implies a C^k-small perturbation of the local solution v), but also to deform slightly the hypersurface Σ in order to make it a real analytic submanifold, which is a convenient setting if one wants to apply the Cauchy-Kowalewski theorem. Let us elaborate on this point.

Given a compact submanifold Σ of codimension $m < n$ (orientable and without boundary), we claim that there exists a C^∞ diffeomorphism $\Phi : \mathbb{R}^n \to \mathbb{R}^n$ such that $\Phi(\Sigma)$ is a real analytic submanifold of $\mathbb{R}^n$ and, Φ is C^k-close to the identity. Indeed, for simplicity, and because it covers all the cases that we shall consider in these notes, we can assume that the normal bundle of Σ is trivial, which amounts to saying that there exists a smooth map $f : \mathbb{R}^n \to \mathbb{R}^m$ such that Σ is a connected component of a regular level set of f. Whitney's approximation theorem then enables us to take a C^ω map $g : \mathbb{R}^n \to \mathbb{R}^m$ which is C^k-close to f. We can now apply Theorem 3.14 to derive the existence of the desired smooth diffeomorphisms Φ, as we wanted to show.

Although in these notes we will only deal with compact level sets, let us briefly comment on the noncompact case, which arises when studying the level sets of harmonic functions. The problem of the structural stability is much subtler. In this case, the condition $\nabla v|_\Sigma \neq 0$ is not enough for the stability. Instead, the function v must satisfy that the infimum of $|\nabla v|$ is positive in a saturated neighborhood of Σ (roughly speaking, in $v^{-1}((c - \epsilon, c + \epsilon))$), and a C^k bound. This extension of the classical Thom's isotopy theorem was proved in [15].

3.3.2 A Runge-Type Global Approximation Theorem for the Helmholtz Equation with Optimal Decay at Infinity

The last ingredient in the strategy explained above is an approximation theorem that allows us to approximate the local solution v by a global solution in a certain neighborhood of the hypersurface Σ. This approximation must be small in the C^k norm in order to apply Thom's isotopy theorem. We say that this result is a Runge-type theorem because it generalizes the classical theorem of Runge in complex analysis ensuring that a holomorphic function defined in a compact set K, whose complement $\mathbb{C} \backslash K$ is connected, can be approximated (in the C^k norm) by a global holomorphic function. The precise statement regarding solutions to the Helmholtz equation is the following. As usual, we say that a function satisfies a PDE in a compact set K if it satisfies the equation in an open set containing K.

Theorem 3.15 *Let v be a solution to the Helmholtz equation $\Delta v + v = 0$ in a compact set $K \subset \mathbb{R}^n$, not necessarily connected. Suppose that the complement $\mathbb{R}^n \backslash K$ is connected. Then there is a function u, satisfying the Helmholtz equation $\Delta u + u = 0$ in $\mathbb{R}^n$, which falls off at infinity as $|D^\alpha u(x)| < C_\alpha |x|^{\frac{1-n}{2}}$ for any multiindex α, and approximates the local solution v in the C^k norm as $\|u - v\|_{C^k(K)} < \delta$. Here δ is any positive constant.*

In the proof of this theorem we shall make use of the following function:

$$G(x) := \beta \, |x|^{1-\frac{n}{2}} \, Y_{\frac{n}{2}-1}(|x|), \tag{3.8}$$

where $Y_{\frac{n}{2}-1}$ denotes the Bessel function of the second kind and we have set

$$\beta := \frac{2^{1-\frac{n}{2}}\pi}{|\mathbb{S}^{n-1}| \, \Gamma(\frac{n}{2} - 1)},$$

with $|\mathbb{S}^{n-1}|$ the area of the unit $(n-1)$-sphere and Γ the Gamma function. A simple computation in spherical coordinates shows that $\Delta G + G = 0$ everywhere except at the origin and the asymptotics for Bessel functions shows that

$$G(x) = -\frac{1}{|\mathbb{S}^{n-1}| \, |x|^{n-2}} + O(|x|^{3-n})$$

as $x \to 0$. It then follows that G is a Green's function (with pole at the origin) for the Helmholtz equation, so if f is a Schwartz function on $\mathbb{R}^n$, then the convolution $G * f$ satisfies

$$\Delta(G * f) + G * f = f.$$

Proof (Proof of Theorem 3.15) Let N be an open neighborhood of K where v satisfies the Helmholtz equation $\Delta v + v = 0$. Our goal is to construct a solution u of the Helmholtz equation in $\mathbb{R}^n$ that approximates v in the set K. To this end, let us take a smooth function $\chi : \mathbb{R}^n \to \mathbb{R}$ that is equal to 1 in a narrow neighborhood of K and is identically zero outside N.

We can now define a smooth function u_1 on $\mathbb{R}^n$ by setting $u_1 := \chi v$. Here we are assuming that $u_1 := 0$ outside N.

Let us now write

$$u_1 = u_1' + h$$

with

$$u_1'(x) = \int_{\mathbb{R}^n} G(x - y)\, f(y)\, dy, \tag{3.9}$$

where f is the compactly supported function $f := \Delta u_1 + u_1$ and G is the Green's function (3.8). By construction, h satisfies the homogeneous Helmholtz equation

$$\Delta h + h = 0.$$

The support of the function f is obviously contained in the open set $N\backslash K$. Therefore, an easy continuity argument ensures that one can approximate the integral (3.9) uniformly in the compact set K by a finite Riemann sum of the form

$$u_2(x) := \sum_{n=1}^{M} c_n\, G(x - x_n). \tag{3.10}$$

Specifically, it is standard that for any $\delta > 0$ there is a large integer M, real numbers c_n and points $x_n \in N\backslash K$ such that the finite sum (3.10) satisfies

$$\|u_1' - u_2\|_{C^0(K)} < \delta. \tag{3.11}$$

Let us now take a large ball B_R containing the closure of the set N. We shall next show that there is a finite number of points $\{x_n'\}_{n=1}^{M'}$ in $\mathbb{R}^n\backslash \overline{B_R}$ and constants c_n' such that the finite linear combination

$$u_3(x) := \sum_{n=1}^{M'} c_n'\, G(x - x_n') \tag{3.12}$$

approximates the function u_2 uniformly in K:

$$\|u_2 - u_3\|_{C^0(K)} < \delta. \tag{3.13}$$

Here δ is the same arbitrarily small constant as above.

Consider the space $\mathscr{V}$ of all finite linear combinations of the form (3.12) where x_n' can be any point in $\mathbb{R}^n\backslash \overline{B_R}$ and the constants c_n' take arbitrary values. Restricting these functions to the set K, $\mathscr{V}$ can be regarded as a subspace of the Banach space $C^0(K)$ of continuous functions on K.

By the Riesz–Markov theorem, the dual of $C^0(K)$ is the space $\mathscr{M}(K)$ of the finite signed Borel measures on $\mathbb{R}^n$ whose support is contained in the set K. Let us take any measure $\mu \in \mathscr{M}(K)$ such that $\int_{\mathbb{R}^n} f\, d\mu = 0$ for all $f \in \mathscr{V}$. We now define a function $F \in L^1_{\text{loc}}(\mathbb{R}^n)$ as

$$F(x) := \int_{\mathbb{R}^d} G(x - x')\, d\mu(x'),$$

so that F satisfies, in the sense of distributions, the equation

$$\Delta F + F = \mu \,.$$

Notice that F is identically zero on $\mathbb{R}^n \setminus \overline{B_R}$ by the definition of the measure μ and that F satisfies the elliptic equation

$$\Delta F + F = 0$$

in $\mathbb{R}^n \setminus K$, so F is analytic in this set. Hence, since $\mathbb{R}^n \setminus K$ is connected by assumption, and contains the set $\mathbb{R}^n \setminus B_R$, by analyticity the function F must vanish on the complement of K. It then follows that the measure μ also annihilates the function $G(\cdot - y)$ with $y \notin K$ because

$$0 = F(y) = \int_{\mathbb{R}^n} G(y - x') \, d\mu(x') \,.$$

Therefore

$$\int_{\mathbb{R}^n} u_2 \, d\mu = 0 \,,$$

which implies that u_2 can be uniformly approximated on K by elements of the subspace $\mathcal{V}$, due to the Hahn–Banach theorem. Accordingly, there is a finite set of points $\{x'_n\}_{n=1}^{M'}$ in $\mathbb{R}^n \setminus \overline{B_R}$ and reals c'_n such that the function u_3 defined by (3.12) satisfies the estimate (3.13).

To complete the proof of the theorem, notice that the function

$$u_4 := u_3 + h$$

satisfies the equation

$$\Delta u_4 + u_4 = 0 \tag{3.14}$$

in the ball B_R, whose interior contains K. Let us take hyperspherical coordinates $r := |x|$ and $\omega := x/|x| \in \mathbb{S}^{n-1}$ in B_R. We can expand the function u_4 (with respect to the angular variables) in a series of spherical harmonics as

$$u_4 = \sum_{l=0}^{\infty} \sum_{m \in I_l} R_{lm}(r) \, Y_{lm}(\omega) \,,$$

where the functions Y_{lm} form a basis of orthonormal spherical harmonics on $\mathbb{S}^{n-1}$ and I_l is a finite set that depends on l and whose explicit expression will not be needed here. Obviously, this series converges in $L^2(B_R)$. Notice that the coefficient

$R_{lm}(r)$ is precisely

$$R_{lm}(r) = \int_{\mathbb{S}^{n-1}} u_4(r\omega)\, Y_{lm}(\omega)\, d\sigma(\omega),$$

where $d\sigma$ is the standard measure on the unit $(n-1)$-sphere, so it is a smooth function. Using Eq. (3.14) it is easy to check that $R_{lm}(r)$ satisfies the ODE

$$R''_{lm} + \frac{n-1}{r} R'_{lm} + \left(1 - \frac{\mu_l}{r^2}\right) R_{lm} = 0,$$

where μ_l is the eigenvalue of the spherical harmonic Y_{lm} (which does not depend on m). Solving this ODE and using that R_{lm} is smooth we infer that

$$R_{lm}(r) = c_{lm}\, j_l(r)$$

where j_l denotes an n-dimensional hyperspherical Bessel function, and c_{lm} is a constant. therefore, we conclude that u_4 can be written in the ball as a series of the form

$$u_4 = \sum_{l=0}^{\infty} \sum_{m \in I_l} c_{lm}\, j_l(r)\, Y_{lm}(\omega).$$

The convergence in $L^2(B_R)$ of this series implies that for any $\delta > 0$ there is an integer l_0 such that the finite sum

$$u := \sum_{l=0}^{l_0} \sum_{m \in I_l} c_{lm}\, j_l(r)\, Y_{lm}(\omega) \tag{3.15}$$

approximates the function u_4 in an L^2 sense:

$$\|u - u_4\|_{L^2(B_R)} < \delta. \tag{3.16}$$

By the properties of Bessel functions, u is smooth in $\mathbb{R}^n$, falls off as

$$|D^\alpha u(x)| < C_\alpha |x|^{\frac{1-n}{2}}$$

at infinity for any multiindex α and satisfies the Helmholtz equation

$$\Delta u + u = 0 \tag{3.17}$$

in the whole space.

Given any $R' < R$ large enough for the set K to be contained in the ball $B_{R'}$, standard elliptic estimates allow us to pass from the L^2 bound (3.16) to a uniform estimate

$$\|u - u_4\|_{C^0(B_{R'})} < C\delta.$$

From this inequality and the bounds (3.11) and (3.13) we infer

$$\|u - v\|_{C^0(K)} = \|u - u_1\|_{C^0(K)} < C\delta. \tag{3.18}$$

Moreover, since v satisfies the Helmholtz equation in a neighborhood of the compact set K, standard elliptic estimates again imply that the uniform estimate (3.18) can be promoted to the C^k bound

$$\|u - v\|_{C^k(K)} < C\delta, \tag{3.19}$$

as we wanted to show.

Remark 3.16 The decay $|D^\alpha u(x)| < C_\alpha |x|^{\frac{1-n}{2}}$ as $|x| \to \infty$ is sharp. This is an improvement with respect to other Runge-type global approximation theorems for elliptic PDEs, where the growth at infinity of the global solutions is not controlled [5]. In particular, there are no finite energy (i.e. square-integrable) solutions to the Helmholtz equation.

3.4 Monochromatic Waves: Nodal Sets of Solutions to the Helmholtz Equation

The solutions to the Helmholtz equation in Euclidean space are usually called monochromatic waves. In this section we use the strategy and techniques introduced in Sect. 3.3 to answer affirmatively Question 3.12. Actually, we will prove a result that is considerably more general than this question, as it applies to an arbitrary number of hypersurfaces. There is, however, a topological condition that we must impose on these hypersurfaces, which is described in the following

Definition 3.17 Let $\Sigma_1, \ldots, \Sigma_d$ be compact smooth orientable codimension one submanifolds without boundary of $\mathbb{R}^n$. We will say that they are *not linked* if there are d pairwise disjoint contractible sets $S_1, \ldots, S_d$ such that each submanifold Σ_i is contained in S_i.

We are now ready to state and prove the main result of this section. Notice that the proof of the theorem provides a satisfactory description of the structure of the diffeomorphism Φ, as noted in Remark 3.19 below. Observe that, of course, for $d = 1$ the condition that the hypersurface be not linked is empty, as it is satisfied trivially. Let us recall that a (connected component of a) level set $\Sigma \subset u^{-1}(0)$

of a function u is *structurally stable* if there is some $\epsilon > 0$ such that if u' is a function with $\|u - u'\|_{C^k} < \epsilon$, then the level set $u'^{-1}(0)$ has a connected component diffeomorphic to Σ (the diffeomorphism is C^k-close to the identity).

Theorem 3.18 *Let $\Sigma_1, \ldots, \Sigma_d$ be compact orientable codimension one submanifolds without boundary of $\mathbb{R}^n$ that are not linked. Then there is a monochromatic wave, i.e. a function u satisfying the Helmholtz equation*

$$\Delta u + u = 0$$

in $\mathbb{R}^n$, and a diffeomorphism Φ of $\mathbb{R}^n$ such that $\Phi(\Sigma_1), \cdots, \Phi(\Sigma_d)$ are structurally stable connected components of the zero set $u^{-1}(0)$. Furthermore, u falls off at infinity as $|D^\alpha u(x)| < C_\alpha |x|^{\frac{1-n}{2}}$ for any multiindex α.

Proof As shown in Sect. 3.3.1, an easy application of Whitney's approximation theorem ensures that, by perturbing the hypersurfaces Σ_i a little if necessary, we can assume that Σ_i is a real analytic hypersurface of $\mathbb{R}^n$. The fact that the hypersurfaces are not linked allow us now to rescale and translate them so that the (unique) precompact domains Ω_i that are bounded by each rescaled and translated real-analytic hypersurface, which we will call $\Sigma'_i := \partial \Omega_i$, are pairwise disjoint and their first Dirichlet eigenvalue $\lambda_1(\Omega_i)$ is 1. It is obvious that there is a diffeomorphism $\Phi' : \mathbb{R}^n \to \mathbb{R}^n$ such that $\Phi'(\Sigma_i) = \Sigma'_i$.

The first eigenvalue $\lambda_1(\Omega_i)$ is always simple, so there is a unique eigenfunction ψ_i, modulo a multiplicative constant, that satisfies the eigenvalue equation

$$\Delta \psi_i + \psi_i = 0 \quad \text{in } \Omega_i, \qquad \psi_i|_{\Sigma'_i} = 0.$$

We can choose ψ_i so that it is positive in Ω_i.

Hopf's boundary point lemma shows that the gradient of ψ_i does not vanish on Σ'_i:

$$\min_{x \in \Sigma'_i} |\nabla \psi_i(x)| > 0. \tag{3.20}$$

Furthermore, as the hypersurface Σ'_i is analytic, it is standard that ψ_i is analytic in an open neighborhood N_i of the closure $\overline{\Omega}_i$.

Our goal is to construct a solution u of the Helmholtz equation in $\mathbb{R}^n$ that approximates each function ψ_i in the set Ω_i. To this end, we define the compact set K

$$K := \bigcup_{i=1}^{d} \overline{\Omega}_i,$$

and the function v in K as

$$v = \psi_i \quad \text{in} \quad \overline{\Omega}_i.$$

Since v satisfies the Helmholtz equation $\Delta v + v = 0$ in the neighborhood

$$N := \bigcup_{i=1}^{d} N_i$$

of K, and the complement of K is connected, the global approximation Theorem 3.15 implies that there is a function u satisfying the Helmholtz equation $\Delta u + u = 0$ in $\mathbb{R}^n$ such that

$$\|u - v\|_{C^k(K)} < \delta. \tag{3.21}$$

Finally, since $\Sigma'_1 \cup \cdots \cup \Sigma'_d$ is a union of components of the nodal set of v and the gradient of v does not vanish on these hypersurfaces by (3.20), the estimate (3.21) and a direct application of Thom's isotopy Theorem 3.14 imply that there is a diffeomorphism Φ'' of $\mathbb{R}^n$ such that

$$\Phi''(\Sigma'_1 \cup \cdots \cup \Sigma'_d) \tag{3.22}$$

is a union of components of the zero set $u^{-1}(0)$. Moreover, the diffeomorphism Φ'' is C^k-close to the identity. Then, the diffeomorphism $\Phi := \Phi'' \circ \Phi'$ transforms the hypersurfaces Σ_i onto a union of connected components of the nodal set of u, as we wanted to prove.

Remark 3.19 It follows from the proof that there is a diffeomorphism Φ', which is the composition of translations and rescalings, and a diffeomorphism Φ'' with $\|\Phi'' - \mathrm{id}\|_{C^k(\mathbb{R}^n)}$ arbitrarily small such that

$$\Phi(\Sigma_i) = (\Phi'' \circ \Phi')(\Sigma_i).$$

In particular, if $d = 1$ the diffeomorphism Φ can be assumed to be an ϵ-rescaling (i.e. the composition of a rescaling with a diffeomorphism ϵ-close to the identity).

3.5 Emergence of Knotted Structures in High-Energy Eigenfunctions: Berry's Conjecture

In this section we will explore the emergence of topological structures in spectral theory. In contrast to monochromatic waves, which are not squared-integrable, here we will be interested in high-energy L^2-eigenfunctions of Schrödinger operators in $\mathbb{R}^n$. The man difficulty in this setting is that we lack of a global approximation theorem like Theorem 3.15. This is not just a technical issue, but a fundamental obstruction in any Runge-type global approximation theorem: for a fixed eigenvalue, the space of L^2-eigenfunctions of a Schrödinger operator is too small. However, as

Fig. 3.3 Berry's conjecture involves showing that there are high-energy eigenfunctions ψ of the harmonic oscillator realizing an arbitrary link (e.g. the trefoil knot or the Borromean rings depicted above) in their nodal set $\psi^{-1}(0)$

we shall see, the analysis we made in Sect. 3.4 of the nodal set of monochromatic waves will be key in this setting because the Helmholtz equation arises, as a rescaled limit, from any well-behaved eigenvalue problem.

To be precise in our analysis of high energy eigenfunctions, we will discuss a conjecture of Sir Michael Berry [4] concerning the nodal set of eigenfunctions of Schrödinger operators in $\mathbb{R}^3$. In his paper, Berry constructs eigenfunctions of the hydrogen atom in $\mathbb{R}^3$ whose nodal set contains a certain torus knot as a connected component (we recall that a knot is a closed curve smoothly embedded in $\mathbb{R}^3$). He then raises the question as to whether there exist eigenfunctions of a quantum system whose nodal set has components with higher order linking, as in the case of the Borromean rings, see Fig. 3.3. It is worth mentioning [4, 27] that a physical motivation to study the nodal set of a quantum system is that it is the locus of destructive interference of the wave function. It is related to the existence of singularities (often called dislocations) of the phase $\mathrm{Im}(\log \psi)$ and of vortices in the current field $\mathrm{Im}(\overline{\psi} \nabla \psi)$. The existence of knotted structures of this type in physical models, especially in optics and in fluid mechanics, has recently attracted considerable attention, both from the theoretical [13, 16] and experimental [12, 28] viewpoints.

One can indeed solve these problems of Berry by showing [22] that any finite link (i.e., any finite collection of pairwise disjoint knots) can be realized as a collection of connected components of the nodal set of a high-energy eigenfunction of the harmonic oscillator. We recall that the eigenfunctions of the harmonic oscillator are the H^1 functions ψ satisfying the equation

$$- \Delta \psi + |x|^2 \psi = \lambda \psi \tag{3.23}$$

in $\mathbb{R}^3$. It is well-known that the eigenvalues are of the form

$$\lambda = 2N + 3 \,,$$

with N a nonnegative integer, and that the multiplicity of the corresponding eigenspace is $\frac{1}{2}(N+1)(N+2)$.

The idea of using high-energy eigenfunctions to construct solutions exhibiting complicated topological structures was first introduced in [21] where we constructed high-energy Beltrami fields on some compact manifolds with vortex lines and vortex tubes of arbitrary link type. The result that ensures that there are high-energy eigenfunctions of the harmonic oscillator whose nodal set contains a knot of any fixed topology can be stated as follows:

Theorem 3.20 *Let L be any finite link in $\mathbb{R}^3$. Then, for any large enough N there is a complex-valued eigenfunction ψ of the harmonic oscillator with eigenvalue $\lambda = 2N+3$ whose nodal set $\psi^{-1}(0)$ has a subset of connected components diffeomorphic to L.*

Remark 3.21 It follows from the proof that the linked components of the nodal set that we construct are actually contained in a small ball whose radius is of order $\lambda^{-1/2}$. These components are structurally stable in the sense that any small enough perturbation of the corresponding eigenfunction (in the C^k norm with $k \geqslant 1$) still has connected components in the nodal set that are diffeomorphic to the knot or link under consideration.

Proof (Proof of Theorem 3.20) The key idea is that, following the same arguments as in the proof of Theorem 3.18, one can show that there are complex-valued solutions to the Helmholtz equation

$$\Delta\varphi + \varphi = 0$$

in $\mathbb{R}^3$ such that the link L is a union of connected components of the nodal set $\varphi^{-1}(0)$, up to a diffeomorphism. This is pertinent to the study of the eigenvalues of the harmonic oscillator because, in balls of radius $\lambda^{-1/2}$, the high-energy asymptotics of the eigenfunctions are determined by the Helmholtz equation. Heuristically, one can understand why this is true by introducing the rescaled variable $\tilde{x} := \lambda^{1/2} x$, in terms of which Eq. (3.23) is read as

$$\Delta_{\tilde{x}}\psi + \psi = \frac{|\tilde{x}|^2 \psi}{\lambda^2}.$$

The way to make this precise is by computing the high-order asymptotics of the Laguerre polynomials, which govern the radial part of the eigenfunctions of the harmonic oscillator. Going over the fine details we will see that the accidental degeneracy of the eigenvalues of the harmonic oscillator is an essential ingredient of the proof too, essentially because it ensures the existence of families of isoenergetic eigenfunctions with a rich behavior in the angular variables.

Since the proof is not overly technical, we shall next give an indication of how the details can be carried out. Let us begin by fixing an orthogonal basis of eigenfunctions associated with the harmonic oscillator Hamiltonian, which we will write in spherical coordinates as

$$\psi_{klm} := e^{-\frac{r^2}{2}} r^l L_k^{l+\frac{1}{2}}(r^2) Y_{lm}(\theta, \phi). \tag{3.24}$$

Here we are using the standard notation for the Laguerre polynomials and the spherical harmonics, the indices of the eigenfunctions range over the set

$$k \geqslant 0, \qquad l \geqslant 0, \qquad -l \leqslant m \leqslant l$$

and the eigenvalue corresponding to ψ_{klm} is

$$\lambda_{kl} := 4k + 2l + 3.$$

Notice that the eigenvalue is independent of m.

It is classical that the behavior of the eigenfunction ψ_{klm} for large values of k is

$$\psi_{klm}(x) = A_{kl} \left[j_l(\sqrt{\lambda_{kl}}\, r) + O(\tfrac{1}{k}) \right] Y_{lm}(\theta, \phi), \tag{3.25}$$

where j_l is the spherical Bessel function and A_{kl} are constants. A useful fact about this expansion is that it only involves Bessel functions, which are solutions to the Helmholtz equation. In fact, the global solution we obtained in the proof of the approximation Theorem 3.15 for the Helmholtz equation is given by a finite series of spherical Bessel functions and spherical harmonics, cf. Eq. (3.15). A modification of the proof of Theorem 3.18 then allows us to construct complex-valued solutions to the Helmholtz equation in $\mathbb{R}^3$ whose zero set has a set of connected components diffeomorphic to the link L. An important additional technical property is that one can choose this global solution to be even:

Lemma 3.22 *There are finitely many complex numbers c_{lm} such that the complex-valued function (which is a solution to the Helmholtz equation in $\mathbb{R}^3$)*

$$\varphi := \sum_{l=0}^{l_0} \sum_{m=-l}^{l} c_{lm}\, j_l(r)\, Y_{lm}(\theta, \phi)$$

has the following properties:

(i) The zero set $\varphi^{-1}(0)$ has a union of connected components diffeomorphic to L which is structurally stable under C^1-small perturbations of φ.
(ii) The function φ is even, so $c_{lm} = 0$ for all odd l.

Proof The proof of this lemma is analogous to the proof of Theorem 3.18, the main modification being the way we construct the local solution of the Helmholtz equation in a neighborhood of the link L. The idea is the following (for simplicity we shall assume that L has one connected component). It is not hard to prove that, up to diffeomorphism, L is the transverse intersection of two real analytic surfaces

Σ_1 and Σ_2. We then can consider the following Cauchy problems, with $j = 1, 2$:

$$\Delta v_j + v_j = 0, \qquad v_j|_{\Sigma_j} = 0, \qquad \partial_\nu v_j|_{\Sigma_j} = 1.$$

Here ∂_ν denotes a normal derivative at the corresponding surface. The Cauchy–Kowalewski theorem then grants the existence of solutions v_j to this Cauchy problem in the closure of small neighborhoods V_j of each surface Σ_j. Now we take the tubular neighborhood $V := V_1 \cap V_2$ of L and define a complex-valued function $\hat\varphi$ on the set V as

$$\hat\varphi := v_1 + i v_2.$$

It is clear from the construction that $\hat\varphi$ satisfies the Helmholtz equation $\Delta\hat\varphi + \hat\varphi = 0$ in V and its nodal set is precisely L. Moreover, the intersection of the zero sets of the real and imaginary parts of $\hat\varphi$ on L is transverse, i.e.,

$$\mathrm{rank}(\nabla\,\mathrm{Re}\,\hat\varphi(x),\, \nabla\,\mathrm{Im}\,\hat\varphi(x)) = 2$$

for all $x \in L$. Thom's isotopy Theorem 3.14 then implies that the nodal set L of $\hat\varphi$ is structurally stable under C^1-small perturbations. The lemma then follows applying Theorem 3.15 to the real and imaginary parts of $\hat\varphi$ to obtain a complex-valued solution φ of the Helmholtz equation in $\mathbb{R}^3$, of the type claimed in the statement, approximating $\hat\varphi$ in the set V. A minor modification of this method using the mirror map $x \mapsto -x$ allows us to construct a function φ that is even.

To exploit this lemma, let us take a large integer $\widehat{k}$ that will be fixed later. For each even integer l smaller than $2\widehat{k}$ we set

$$\widehat{k}_l := \widehat{k} - \frac{l}{2}, \tag{3.26}$$

so that the eigenvalue

$$\lambda := \lambda_{\widehat{k}_l} = 4\widehat{k} + 3 \tag{3.27}$$

does not depend on the choice of l. The desired eigenfunction ψ of the harmonic oscillator can then be derived from the function φ constructed in Lemma 3.22 by setting

$$\psi := \sum_{l=0}^{l_0} \sum_{m=-l}^{l} \frac{c_{lm}}{A_{\widehat{k}_l}} \psi_{\widehat{k}_l l m}$$

for a large enough number $\widehat{k}$. Notice that, by construction, ψ is a smooth complex-valued eigenfunction of the harmonic oscillator with eigenvalue $\lambda = 4\widehat{k} + 3$, and that we have used that $c_{lm} = 0$ for odd l because the number $\widehat{k}_l$ is an integer only for even l.

Using the asymptotics (3.25) it is not hard to show that for any $\delta > 0$ one can choose $\widetilde{k}$ large enough so that

$$\left\| \psi\left(\frac{\cdot}{\sqrt{\lambda}}\right) - \varphi(\cdot) \right\|_{C^1(B)} < \delta . \tag{3.28}$$

Lemma 3.22 then ensures that, if δ is small enough, the function $\psi(\cdot/\sqrt{\lambda})$ has a collection of connected components in its nodal set $\{\psi(\cdot/\sqrt{\lambda}) = 0\}$ diffeomorphic to L, so the theorem follows.

It is worth emphasizing that the key point of the proof is not exactly that the eigenvalue is very high, but rather that it is very degenerate and that the corresponding eigenfunctions have a very rich behavior in the angular variables. Consequently, one can prove a similar result for highly excited states of the hydrogen atom [23], which was the original context that Berry considered. We recall that the eigenfunctions of the hydrogen atom satisfy the equation

$$\left(\Delta + \frac{2}{|x|} + \lambda \right) \psi = 0, \tag{3.29}$$

in $\mathbb{R}^3$. The eigenvalues are given by

$$\lambda_n := -\frac{1}{n^2} ,$$

where n is a positive integer, and the corresponding eigenfunctions are given, with $0 \leqslant l \leqslant n - 1$ and $-l \leqslant m \leqslant l$, by

$$\psi_{nlm} := e^{-r/n} r^l L_{n-l-1}^{2l+1}\left(\frac{2r}{n}\right) Y_{lm}(\theta, \phi) ,$$

where L_{ν}^{α} are the Laguerre polynomials.

The result for the hydrogen atom can be stated as follows:

Theorem 3.23 *Let L be any finite link in $\mathbb{R}^3$. Then, for any Coulomb eigenvalue with energy λ_n close enough to zero, there exist a complex-valued eigenfunction ψ of energy λ_n whose nodal set $\psi^{-1}(0)$ has a union of connected components diffeomorphic to L.*

A key difficulty here, which does not appear in the case of the harmonic oscillator, concerns estimates for the Green's function of the Coulomb Hamiltonian with zero energy, $\Delta + 2/|x|$.

3.6 The Linear Regime of Nonlinear Equations: Nodal Sets of the Allen–Cahn Equation

While in Sect. 3.4 we decided to restrict our attention to monochromatic waves on $\mathbb{R}^n$ (solutions to the Helmholtz equation) to keep statements as simple as possible, it is clear that the results can be extended to significantly wider classes of equations. Specifically, in [15] it is discussed to which extent these results remain valid for more general second-order linear elliptic PDEs. Our concern in this section will be to show how these methods can be used to study nonlinear equations, as long as one stays in the linear regime.

To illustrate the idea, we will consider the paradigmatic case of the Allen–Cahn equation in $\mathbb{R}^n$:

$$\Delta u + u - u^3 = 0.$$

In this context, the basic question that we will discuss in this section is the following:

Question 3.24 Let Σ be a codimension one compact submanifold of $\mathbb{R}^n$ (without boundary and orientable). Is there a solution to the Allen–Cahn equation on $\mathbb{R}^n$ with (a connected component of) a level set diffeomorphic to Σ?

Before getting into the heart of the matter, it is worth recalling that the study of the level sets of solutions to the Allen–Cahn equation has attracted much attention, especially due to De Giorgi's 1978 conjecture that all the level sets of any solution to the Allen–Cahn equation in $\mathbb{R}^n$ that is monotone in one direction must be hyperplanes for $n \leqslant 8$. This is a natural counterpart of the Bernstein problem for minimal hypersurfaces, which asserts that any minimal graph in $\mathbb{R}^n$ must be a hyperplane provided that $n \leqslant 8$. Ghoussoub–Ghi and Ambrosio–Cabré proved De Giorgi's conjecture for $n = 2, 3$ [3, 25], and the work of Savin [32] showed that it is also true for $4 \leqslant n \leqslant 8$ under a natural technical assumption. Del Pino, Kowalczyk and Wei [9] employed the Bombieri–De Giorgi–Giusti hypersurface to show that the statement of De Giorgi's conjecture does not hold for $n \geqslant 9$.

The analysis and possible classification of bounded entire solutions to the Allen–Cahn equation is an important open problem where the Morse index of the solution u (that is, the maximal dimension of a vector space $V \subset C_0^\infty(\mathbb{R}^n)$ such that

$$\int_{\mathbb{R}^n} \left(|\nabla v|^2 - v^2 + 3u^2 v^2 \right) dx < 0$$

for all nonzero $v \in V$) plays a key role. Generally speaking, it is expected [10, 11] that the condition that the Morse index of the solution be finite should play a similar role as the finite total curvature assumption in the study of minimal hypersurfaces in Euclidean spaces. In particular, it is well known that there are many infinite-index solutions to the Allen–Cahn equation [6, 8], and this abundance of solutions should translate into a wealth of possible level sets.

We shall next show how one can use ideas related to the matters discussed in Sects. 3.3 and 3.4 to explore the flexibility of bounded entire solutions to the Allen–Cahn equation of infinite index, proving, in particular, a positive answer to Question 3.24 [19]:

Theorem 3.25 *Let Σ be any compact orientable codimension one submanifold without boundary of $\mathbb{R}^n$, with $n \geqslant 4$. Then there is an entire solution u of the Allen–Cahn equation in $\mathbb{R}^n$, which falls off at infinity as $|u(x)| < C|x|^{\frac{1-n}{2}}$, such that its nodal set $u^{-1}(0)$ has a structurally stable connected component diffeomorphic to Σ.*

Proof (Sketch of the Proof) The result hinges on two key lemmas. The first one concerns the linearization of the Allen–Cahn equation at 0, that is, the Helmholtz equation. Specifically, in the particular case that $d = 1$ in Theorem 3.18, we have the following:

Lemma 3.26 *There is a function w satisfying the Helmholtz equation*

$$\Delta w + w = 0$$

in $\mathbb{R}^n$ and a diffeomorphism Ψ of $\mathbb{R}^n$ such that $\Psi(\Sigma)$ is a structurally stable connected component of the zero set $w^{-1}(0)$. Furthermore, w falls off at infinity as $|D^j w(x)| < C_j (1 + |x|)^{\frac{1-n}{2}}$ for any j.

To pass from the Helmholtz equation to the Allen–Cahn equation one resorts to an iterative procedure. To set the iteration we shall use the fundamental solution G of the Helmholtz equation introduced in Eq. (3.8).

Using the function G, the iterative scheme can then be assumed to be

$$u_0 := \delta w,$$

$$u_{j+1} := \delta w + G * (u_j^3), \tag{3.30}$$

where w is the above solution to the Helmholtz equation and δ is some small positive constant that will be fixed later. One can prove that, if δ is small enough, then u_j converges as $j \to \infty$ to some function u in the weighted space $C^k_{\frac{n-1}{2}}(\mathbb{R}^n)$ of functions whose norm

$$\|u\|_{k,\frac{n-1}{2}} := \max_{|\alpha| \leqslant k} \sup_{x \in \mathbb{R}^n} \left| (1 + |x|)^{\frac{n-1}{2}} \partial^\alpha u(x) \right|$$

is finite. Here $k \geqslant 2$ is any integer. The function u must then satisfy

$$u = \delta w + G * (u^3),$$

so in particular we infer that

$$\Delta u + u - u^3 = 0$$

by applying the Helmholtz operator to the above identity.

To prove the convergence of the iteration, one needs to use the sharp fall off of the function w at infinity and the following lemma, which provides a weighted estimate for convolutions with the fundamental solution G:

Lemma 3.27 *For any* $v \in C^k_{\frac{n-1}{2}}(\mathbb{R}^n)$, *one has the estimate*

$$\|G * (v^3)\|_{k, \frac{n-1}{2}} \leqslant C\|v\|^3_{k, \frac{n-1}{2}}.$$

Once that one has established the converge of the iteration, the rest of the argument is easy. Indeed, it readily follows that the solution to the Allen–Cahn equation must satisfy the bounds

$$\|u\|_{k, \frac{n-1}{2}} \leqslant C\delta$$

and

$$\left\| w - \frac{u}{\delta} \right\|_{k, \frac{n-1}{2}} = \frac{1}{\delta} \|\delta w - u\|_{k, \frac{n-1}{2}} = \frac{1}{\delta} \|G * (u^3)\|_{k, \frac{n-1}{2}} \leqslant \frac{C}{\delta} \|u\|^3_{k, \frac{n-1}{2}} \leqslant C\delta^2.$$

Since the nodal sets of u and u/δ obviously coincide, a moment's thought shows that the fact that the nodal set $w^{-1}(0)$ has a structurally stable connected component diffeomorphic to Σ implies that, for small enough δ the nodal set $u^{-1}(0)$ must also have a connected component that is diffeomorphic to Σ and structurally stable. The theorem then follows.

Acknowledgements The second named author is very grateful to the CIME Foundation and the organizers of the Summer School "Geometry of PDEs and Related Problems" for their kind invitation to deliver one of the courses. The authors are supported by the ERC Starting Grants 633152 (A.E.) and 335079 (D.P.-S.). This work is supported in part by the ICMAT–Severo Ochoa grant SEV-2015-0554.

References

1. R. Abraham, J. Robbin, *Transversal Mappings and Flows* (Benjamin, New York, 1967)
2. G. Alessandrini, R. Magnanini, The index of isolated critical points and solutions of elliptic equations in the plane. Ann. Sc. Norm. Super. Pisa **19**, 567–589 (1992)
3. L. Ambrosio, X. Cabré, Entire solutions of semilinear elliptic equations in $\mathbb{R}^3$ and a conjecture of De Giorgi. J. Am. Math. Soc. **13**, 725–739 (2000)
4. M. Berry, Knotted zeros in the quantum states of hydrogen. Found. Phys. **31**, 659–667 (2001)
5. F.E. Browder, Approximation by solutions of partial differential equations. Am. J. Math. **84**, 134–160 (1962)
6. X. Cabré, J. Terra, Saddle-shaped solutions of bistable diffusion equations in all of $\mathbb{R}^{2m}$. J. Eur. Math. Soc. **11**, 819–843 (2009)
7. S.Y. Cheng, Eigenfunctions and nodal sets. Comment. Math. Helv. **51**, 43–55 (1976)

8. E.N. Dancer, New solutions of equations on $\mathbb{R}^n$. Ann. Sc. Norm. Super. Pisa **30**, 535–563 (2001)
9. M. del Pino, M. Kowalczyk, J. Wei, On De Giorgi conjecture in dimension $n \geqslant 9$. Ann. Math. **174**, 1485–1569 (2011)
10. M. del Pino, M. Kowalczyk, J. Wei, On De Giorgi's conjecture and beyond. Proc. Natl. Acad. Sci. **109**, 6845–6850 (2012)
11. M. del Pino, M. Kowalczyk, J. Wei, Entire solutions of the Allen–Cahn equation and complete embedded minimal surfaces of finite total curvature in $\mathbb{R}^3$. J. Differ. Geom. **93**, 67–131 (2013)
12. M.R. Dennis, R.P. King, B. Jack, K. O'Holleran, M.J. Padgett, Isolated optical vortex knots. Nat. Phys. **6**, 118–121 (2010)
13. A. Enciso, D. Peralta-Salas, Knots and links in steady solutions of the Euler equation. Ann. Math. **175**, 345–367 (2012)
14. A. Enciso, D. Peralta-Salas, Critical points of Green's functions on complete manifolds. J. Differ. Geom. **92**, 1–29 (2012)
15. A. Enciso, D. Peralta-Salas, Submanifolds that are level sets of solutions to a second-order elliptic PDE. Adv. Math. **249**, 204–249 (2013)
16. A. Enciso, D. Peralta-Salas, Existence of knotted vortex tubes in steady Euler flows. Acta Math. **214**, 61–134 (2015)
17. A. Enciso, D. Peralta-Salas, Eigenfunctions with prescribed nodal sets. J. Differ. Geom. **101**, 197–211 (2015)
18. A. Enciso, D. Peralta-Salas, Critical points and geometric properties of Green's functions on open surfaces. Ann. Mat. Pura Appl. **194**, 881–901 (2015)
19. A. Enciso, D. Peralta-Salas, Bounded solutions to the Allen–Cahn equation with level sets of any compact topology. Anal. PDE **9**, 1433–1446 (2016)
20. A. Enciso, D. Peralta-Salas, S. Steinerberger, Prescribing the nodal set of the first eigenfunction in each conformal class. Int. Math. Res. Notices **2017**, 3322–3349 (2017)
21. A. Enciso, D. Peralta-Salas, F. Torres de Lizaur, Knotted structures in high-energy Beltrami fields on the torus and the sphere. Ann. Sci. Éc. Norm. Supér. **50**, 995–1016 (2017)
22. A. Enciso, D. Hartley, D. Peralta-Salas, A problem of Berry and knotted zeros in the eigenfunctions of the harmonic oscillator. J. Eur. Math. Soc. **20**, 301–314 (2018)
23. A. Enciso, D. Hartley, D. Peralta-Salas, Dislocations of arbitrary topology in Coulomb eigenfunctions. Rev. Mat. Iberoam. (in press)
24. L. Flatto, D.J. Newman, H.S. Shapiro, The level curves of harmonic functions. Trans. Am. Math. Soc. **123**, 425–436 (1966)
25. N. Ghoussoub, C. Gui, On a conjecture of De Giorgi and some related problems. Math. Ann. **311**, 481–491 (1998)
26. R. Hardt, L. Simon, Nodal sets for solutions of elliptic equations. J. Differ. Geom. **30**, 505–522 (1989)
27. L.H. Kauffman, S.J. Lomonaco, Quantum knots. Proc. SPIE **5436**, 268–284 (2004)
28. D. Kleckner, W.T.M. Irvine, Creation and dynamics of knotted vortices. Nat. Phys. **9**, 253–258 (2013)
29. P. Li, L.F. Tam, Symmetric Green's functions on complete manifolds. Am. J. Math. **109**, 1129–1154 (1987)
30. J.C. Maxwell, *A Treatise on Electricity and Magnetism*, vol. I (Oxford University Press, Oxford, 1998)
31. L. Perko, *Differential Equations and Dynamical Systems* (Springer, New York, 2001)
32. O. Savin, Regularity of flat level sets in phase transitions. Ann. Math. **169**, 41–78 (2009)
33. R. Thom, Quelques propriétés globales des variétés différentiables. Comment. Math. Helv. **28**, 17–86 (1954)

Chapter 4
Symmetry Properties for Solutions of Higher-Order Elliptic Boundary Value Problems

Wolfgang Reichel

Abstract Goals of this lecture-series: The lecture series is intended to give a glimpse into different methods that allow to conclude that solutions of higher-order nonlinear boundary value problems have the same symmetries as the underlying domain. In the second-order case this has been a very prominent area of research for the last half-century. For higher-order cases the theory has not yet been developed up to a comparable level. We will show in an exemplary way how different techniques from nonlinear analysis and geometry can be used to answer symmetry questions. In particular, we will discuss symmetry properties of solutions to $(-\Delta)^m u = f(x, u)$ either on $\mathbb{R}^n$ or on balls with additional Dirichlet boundary conditions. We will give examples for three different methods:

- methods based on contraction mapping
- methods based on the moving plane method
- an example of an overdetermined 4th order problem using Newton's inequalities

The first two sections are very much inspired by Gazzola et al. (Polyharmonic boundary value problems. Lecture Notes in Mathematics, vol 1991. Springer, Berlin, 2010. Positivity preserving and nonlinear higher order elliptic equations in bounded domains), where much more extended material on linear and nonlinear polyharmonic boundary values problems can be found. The fourth section makes use of the moving plane method, and references are given in this section. The final section shows how Newton's inequalities may be used to solve a fourth-order overdetermined boundary value problem. This section is the result of joint discussions with V. Ferone, C. Nitsch and C. Trombetti from Univ. di Napoli "Federico II".

W. Reichel (✉)
Institute for Analysis, Department of Mathematics, Karlsruhe Institute of Technology, Karlsruhe, Germany
e-mail: wolfgang.reichel@kit.edu

© Springer Nature Switzerland AG 2018
C. Bianchini et al. (eds.), *Geometry of PDEs and Related Problems*,
Lecture Notes in Mathematics 2220, https://doi.org/10.1007/978-3-319-95186-7_4

4.1 Linear Problems: Weak Solutions, Eigenvalues, Regularity, Green Functions

Suppose $m \in \mathbb{N}$ and let $\Omega \subset \mathbb{R}^n$ be a bounded domain with Lipschitz boundary so that for almost all $x \in \partial\Omega$ the outer unit normal $\nu(x)$ exists. Here we consider the polyharmonic boundary value problem

$$(-\Delta)^m u = f(x) \text{ in } \Omega, \quad u = \frac{\partial}{\partial \nu} u = \ldots = \left(\frac{\partial}{\partial \nu}\right)^{m-1} u = 0 \text{ on } \partial\Omega, \qquad (4.1)$$

where the Laplace operator (Laplacian) is defined by $\Delta := \sum_{i=1}^{n} \frac{\partial^2}{\partial x_i^2}$ and the polyharmonic operator (or iterated Laplacian) is defined recursively as $(-\Delta)^m := (-\Delta) \circ (-\Delta)^{m-1}$ for $m \in \mathbb{N}$, $m \geq 2$.

To formulate solutions concepts for (4.1) we need to define some function spaces. The classical L^p-spaces on a domain $\Omega \subset \mathbb{R}^n$ are denoted by $L^p(\Omega)$ for $1 \leq p \leq \infty$. For derivatives of a function $u : \Omega \to \mathbb{R}$ we use the multi-index notation with a multi-index $\alpha = (\alpha_1, \ldots, \alpha_n) \in \mathbb{N}_0^n$ to define

$$D^\alpha u = \frac{\partial^{|\alpha|}}{\partial_{x_1}^{\alpha_1} \cdots \partial_{x_n}^{\alpha_n}} u \text{ with } |\alpha| = \alpha_1 + \ldots + \alpha_n.$$

Let $C_c^\infty(\Omega)$ be the space of test-functions. For $0 \leq \gamma \leq 1$ and $k \in \mathbb{N}_0$ let

$$C^k(\overline{\Omega}) := \{u : \Omega \to \mathbb{R} : D^\alpha u \text{ exist and have continuous and bounded}$$
$$\text{extensions to } \overline{\Omega} \, \forall \alpha \text{ with } |\alpha| \leq k\}$$
$$C_0^k(\overline{\Omega}) := \{u \in C^k(\overline{\Omega}) : D^\alpha u|_{\partial\Omega} = 0 \, \forall \, \alpha \text{ with } |\alpha| \leq k\}$$
$$C^{k,\gamma}(\overline{\Omega}) := \{u \in C^k(\overline{\Omega}) : [D^\alpha u]_\gamma < \infty \, \forall \alpha \text{ with } |\alpha| = k\}$$

where $[f]_\gamma = \sup_{x,y \in \Omega, x \neq y} \frac{|f(x)-f(y)|}{|x-y|^\gamma}$. The spaces are normed by

$$\|u\|_{C^k} := \max_{|\alpha| \leq k} \|D^\alpha u\|_\infty, \qquad \|u\|_{C^{k,\gamma}} := \|u\|_{C^k} + \max_{|\alpha|=k} [D^\alpha u]_\gamma$$

and $C_0^k(\overline{\Omega})$ is a closed subspace of $C^k(\overline{\Omega})$. For functions $u : \Omega \to \mathbb{R}$ having weak derivatives of order $|\alpha|$ we use the symbol $D^\alpha u : \Omega \to \mathbb{R}$ to denote the weak derivative. For $1 \leq p \leq \infty$ let

$$W^{k,p}(\Omega) = \{u : \Omega \to \mathbb{R} : D^\alpha u \in L^p(\Omega) \, \forall \alpha \text{ with } |\alpha| \leq k\}$$

with norm $\|u\|_{W^{k,p}} := \left(\sum_{|\alpha| \leq k} \|D^\alpha u\|_{L^p}^p\right)^{1/p}$ if $1 \leq p < \infty$ and $\|u\|_{W^{k,\infty}} := \max_{|\alpha| \leq k} \|D^\alpha u\|_\infty$. A closed subspace of $W^{k,p}(\Omega)$ encoding the concept of vanishing boundary values up to order $k - 1$ is defined as

$$W_0^{k,p}(\Omega) = \overline{C_c^\infty(\Omega)}^{\|\cdot\|_{W^{k,p}}}.$$

For bounded domains Ω the spaces $W_0^{k,p}(\Omega)$ can be equipped with the equivalent norm $\|u\|_{W_0^{k,p}} := \left(\sum_{|\alpha|=k} \|D^\alpha u\|_{L^p}^p\right)^{1/p}$ if $1 \le p < \infty$ and $\|u\|_{W_0^{k,\infty}} := \max_{|\alpha|=k} \|D^\alpha u\|_\infty$. This is a consequence of the Poincaré inequality $\|u\|_{L^p} \le C(j, \Omega, p)\|\partial_j u\|_{L^p}$ for $u \in W_0^{1,p}(\Omega)$ and $j = 1, \ldots, n$. In the case $p = 2$ we define the Hilbert spaces $H^k(\Omega) := W^{k,2}(\Omega)$ and $H_0^k(\Omega) := W_0^{k,2}(\Omega)$.

On the Hilbert space $H_0^k(\Omega)$ one can introduce yet another interesting equivalent norm.

Lemma 4.1 *Let $k \in \mathbb{N}$. For $u, v \in H_0^k(\Omega)$ define*

$$\mathscr{D}^k u = \begin{cases} \Delta^r u & \text{if } k = 2r, \\ \nabla\Delta^r u & \text{if } k = 2r + 1, \end{cases} \qquad \mathscr{D}^k u \mathscr{D}^k v = \begin{cases} \Delta^r u \Delta^r v & \text{if } k = 2r, \\ \nabla\Delta^r u \cdot \nabla\Delta^r v & \text{if } k = 2r + 1. \end{cases}$$

Then the scalar product

$$\langle u, v \rangle_{H_0^k} := \int_\Omega \mathscr{D}^k u \, \mathscr{D}^k v \, dx, \quad u, v \in H_0^k(\Omega)$$

generates via $\|u\|_{H_0^k} := \langle u, u \rangle_{H_0^k}^{\frac{1}{2}}$ a norm which is equivalent to $\|\cdot\|_{W_0^{k,2}}$.

Proof Observe that $\langle u, v \rangle_{H_0^k} = \sum_{|\alpha|=k} \int_\Omega D^\alpha u D^\alpha v \, dx$ for all $u, v \in H_0^k(\Omega)$. By density it is enough to verify this for $u, v \in C_c^\infty(\Omega)$. The relation is then a consequence of several integrations by parts. For illustration let us, e.g., consider the case $k = 2$:

$$\sum_{|\alpha|=2} \int_\Omega D^\alpha u D^\alpha v \, dx = \sum_{i,j=1}^n \int_\Omega \frac{\partial^2}{\partial x_i \partial x_j} u \frac{\partial^2}{\partial x_i \partial x_j} v \, dx$$

$$= \sum_{i,j=1}^n \int_\Omega \frac{\partial^2}{\partial x_i \partial x_i} u \frac{\partial^2}{\partial x_j \partial x_j} v \, dx$$

$$= \int_\Omega \Delta u \Delta v \, dx.$$

$\square$

In the following we consider three kinds of solutions: weak, strong and classical.

Definition 4.2 (Weak Solution) Let $f \in L^2(\Omega)$. A function $u \in H_0^m(\Omega)$ is called a weak solution of (4.1) if

$$\int_\Omega \mathscr{D}^m u \, \mathscr{D}^m \phi \, dx = \int_\Omega f \phi \, dx \text{ for all } \phi \in H_0^m(\Omega).$$

Using the Riesz-representation theorem as in [17], Chapter 4.5 and the spectral theorem for compact, symmetric operators as in [11], Chapter 6.5 one can establish the following result in analogy to the case of the Laplacian presented in the preceding two references.

Lemma 4.3 *For every $f \in L^2(\Omega)$ there exists a unique weak solution $u \in H_0^m(\Omega)$ of (4.1) and the solution operator $T : L^2(\Omega) \to L^2(\Omega)$, $Tf := u$ is compact. It has a complete L^2-orthonormal basis $\{\phi_i : i \in \mathbb{N}\}$ of eigenfunctions with corresponding eigenvalues $\{\lambda_i : i \in \mathbb{N}\}$, where the eigenvalues λ_i are ordered*

$$\lambda_1 \leq \lambda_2 \leq \lambda_3 \ldots$$

Within the eigenvalue sequence, values are repeated according to the multiplicity of the eigenvalue.

Definition 4.4 (Strong Solution) Let $f \in L^p(\Omega)$ for some $p \in (1, \infty)$. A function $u \in W^{2m,p}(\Omega) \cap W_0^{m,p}(\Omega)$ is called a strong solution of (4.1) provided the differential equation in (4.1) holds for almost all $x \in \Omega$.

The following result establishes existence and uniqueness of strong solutions to (4.1), cf. Corollary 2.21 in [13].

Lemma 4.5 *Let $1 < p < \infty$ and assume $\partial\Omega \in C^{2m}$. For every $f \in L^p(\Omega)$ there exists a unique strong solution $u \in W^{2m,p}(\Omega) \cap W_0^{m,p}(\Omega)$ of (4.1) and the solution operator $T : L^p(\Omega) \to W^{2m,p}(\Omega) \cap W_0^{m,p}(\Omega)$, $Tf := u$ is bounded.*

Definition 4.6 (Classical Solution) Let $f \in C(\overline{\Omega})$. A function $u \in C^{2m}(\overline{\Omega}) \cap C_0^{m-1}(\overline{\Omega})$ is called a classical solution of (4.1) provided the differential equation in (4.1) holds for all $x \in \Omega$.

The corresponding existence and uniqueness result for classical solutions to (4.1) can be found in Theorem 2.19 in [13].

Lemma 4.7 *Assume $\partial\Omega \in C^{2m,\alpha}$. For every $f \in C^\alpha(\overline{\Omega})$ there exists a unique classical solution $u \in C^{2m,\alpha}(\overline{\Omega}) \cap C_0^m(\overline{\Omega})$ of (4.1) and the solution operator $T : C^\alpha(\overline{\Omega}) \to C^{2m,\alpha}(\overline{\Omega}) \cap C_0^m(\overline{\Omega})$, $Tf := u$ is bounded.*

Definition 4.8 (Positivity Preserving Property) The operator T from Lemma 4.3, Lemma 4.5 or Lemma 4.7 is called positivity preserving if $f \geq 0$ implies $u \geq 0$.

The fact that the solution operator for a ball, i.e., $\Omega = B_R(P) \subset \mathbb{R}^n$, is positivity preserving is deduced from the following formula originally given by Boggio [6]. Details (in particular Lemma 4.10 below) can be found in Section 2.6 in [13].

Definition 4.9 (Green Function for Balls) For $x \in \overline{B_1(0)}$ let $x^* = x/|x|^2$ and define for $x, y \in \overline{B_1(0)}$

$$G(x, y) := k_{m,n}|x - y|^{2m-n} \int_1^{|x| \, |x^*-y|/|x-y|} (t^2 - 1)^{m-1} t^{1-n} \, dt.$$

where $k_{m,n} = \frac{\Gamma(1+n/2)}{n\pi^{n/2} 4^{m-1} \Gamma(m)^2}$.

Lemma 4.10 *Let* $\Omega = B_1(0) \subset \mathbb{R}^n$. *Then for* $1 < p < \infty$ *and* $f \in L^p(B_1(0))$ *the solution operator from Lemma 4.5 is given by*

$$u(x) = (Tf)(x) = \int_{B_1(0)} G(x, y) f(y) \, dy, \quad x \in B_1(0)$$

and hence it is positivity preserving. The same is true for the solution operator of Lemma 4.3 if $f \in L^2(B_1(0))$ *and for the solution operator of Lemma 4.7 if* $f \in C^\alpha(\overline{B_1(0)})$.

4.2   Symmetry, Simplicity and Positivity for First Eigenfunctions

Recall from Lemma 4.3 that $(-\Delta)^m$ with Dirichlet boundary conditions on any bounded domain Ω has a first eigenvalue

$$\lambda_1 = \min_{u \in H_0^m(\Omega) \setminus \{0\}} \frac{\|u\|_{H_0^m}^2}{\|u\|_{L^2}^2}.$$

Three natural questions arise:

1. Is λ_1 simple?
2. If so, do the eigenfunctions corresponding to λ_1 have one sign?
3. If Ω is symmetric, do the eigenfunctions corresponding to λ_1 have the same symmetry as Ω?

In this section we answer all three questions in a positive way when $\Omega = B_1(0)$. Let us first sketch the usual way of answering question (1) in case $m = 1$. Let u be a Dirichlet eigenfunction of $-\Delta$ corresponding to λ_1 and decompose $u = u^+ + u^-$ with the usual pointwise definitions $u^+ = \max\{u, 0\}$ and $u^- = \min\{u, 0\}$. In $H_0^1(\Omega)$ a (nontrivial) fact is that $u^+, u^- \in H_0^1(\Omega)$ and $\nabla u = \nabla u^+ \chi_{\{u>0\}} + \nabla u^- \chi_{\{u<0\}}$ and hence $|\nabla |u|| = |\nabla u|$ a.e. in Ω. Assume, for contradiction, that u is sign-changing, i.e., u^+, u^- are both non-trivial. Thus $|u|, u^+, u^-$ are also minimizers for λ_1. The fact that, e.g., u^+ vanishes on a set of positive measure then contradicts the strong minimum principle. For $m \geq 2$ this argument fails twice.

First, $u \in H_0^m(\Omega)$ does not in general imply that u^+ belongs to $H^m(\Omega)$. Second, the maximum principle in general fails.

The way out for $m \geq 2$ is to decompose $u = u_1 + u_2$ with $u_1 \geq 0$ a.e. in Ω and $u_1 \perp_{H_0^m} u_2$. This relies on an abstract construction due to Moreau [21] as described in Sections 3.1.2 and 3.1.3. in [13].

Definition 4.11 (Dual Cones) Let $(H, \langle \cdot, \cdot \rangle)$ be a real Hilbert space. Let $\mathcal{K} \subset H$ be a cone, i.e., $v \in \mathcal{K}$ implies $tv \in \mathcal{K}$ for all $t \geq 0$. The dual cone $\mathcal{K}^*$ is defined by

$$\mathcal{K}^* = \{w \in H : \langle w, v \rangle \leq 0 \text{ for all } v \in \mathcal{K}\}.$$

Theorem 4.12 ([21]) *Let $(H, \langle \cdot, \cdot \rangle)$ be a real Hilbert space and let $\mathcal{K} \subset H$ be a closed convex cone. Then for every $u \in H$ there exist unique elements $u_1 \in \mathcal{K}$, $u_2 \in \mathcal{K}^*$ such that $u = u_1 + u_2$ with $u_1 \perp u_2$. Moreover for all $u, \tilde{u} \in H$*

$$\|u - \tilde{u}\|^2 \geq \|u_1 - \tilde{u}_1\|^2 + \|u_2 - \tilde{u}_2\|^2. \tag{4.2}$$

Proof One defines u_1 as the (unique) minimizer $\|u - u_1\| = \min_{v \in \mathcal{K}} \|u - v\|$ and $u_2 := u - u_1$. By the cone property and convexity of $\mathcal{K}$ we have $u_1 + tv \in \mathcal{K}$ whenever $t \geq 0$ and $v \in \mathcal{K}$ so that

$$\|u - u_1\|^2 \leq \|u - (u_1 + tv)\|^2 \text{ for all } t \geq 0. \tag{4.3}$$

The quantity on the right-hand side of (4.3) has a minimum at $t = 0$ and thus

$$0 \leq \frac{d}{dt}\|u - (u_1 + tv)\|^2\Big|_{t=0} = -2\langle u - u_1, v \rangle.$$

Hence $\langle u_2, v \rangle \leq 0$ for all $v \in \mathcal{K}$ so that $u_2 \in \mathcal{K}^*$. Moreover, setting $v = u_1$ we see that (4.3) holds for all $t \geq -1$ and thus we can use

$$0 = \frac{d}{dt}\|u - (1 + t)u_1)\|^2\Big|_{t=0} = -2\langle u - u_1, u_1 \rangle$$

so that $u_1 \perp u_2$. Let us finally prove (4.2). Note that for all $u, \tilde{u} \in H$ we have

$$\|u - \tilde{u}\|^2 = \|u_1 - \tilde{u}_1\|^2 + \|u_2 - \tilde{u}_2\|^2 + 2\langle u_1 - \tilde{u}_1, u_2 - \tilde{u}_2 \rangle$$

$$= \|u_1 - \tilde{u}_1\|^2 + \|u_2 - \tilde{u}_2\|^2 - 2\underbrace{\langle u_1, \tilde{u}_2 \rangle}_{\leq 0} - 2\underbrace{\langle \tilde{u}_1, u_2 \rangle}_{\leq 0}$$

$$\geq \|u_1 - \tilde{u}_1\|^2 + \|u_2 - \tilde{u}_2\|^2.$$

This inequality also implies uniqueness of u_1, u_2 with the stated properties. □

Let us briefly discuss the consequences of taking $\mathcal{K} = \{v : v \geq 0\}$ where the functions v are taken from the Hilbert spaces $L^2(\Omega)$, $H_0^1(\Omega)$, $H_0^2(\Omega)$. Clearly in the first case, where $H = L^2(\Omega)$ we get $\mathcal{K} = \{v \in L^2(\Omega) : v \geq 0\}$ and $\mathcal{K}^* = \{w \in L^2(\Omega) : w \leq 0\}$ so that the splitting according to Theorem 4.12 yields $u = u_1 + u_2$ with $u_1 = u^+$, $u_2 = u^-$. In the second case let us equip $H_0^1(\Omega)$ with the scalar product $\langle u, \tilde{u} \rangle_{H_0^1} := \int_\Omega \nabla u \cdot \nabla \tilde{u} \, dx$. Then $\mathcal{K} = \{v \in H_0^1(\Omega) : v \geq 0\}$ and $\mathcal{K}^* = \{w \in H_0^1(\Omega) : w \text{ is weakly sub-harmonic}\}$. Clearly, by the maximum principle, $\mathcal{K}^* \subset -\mathcal{K}$ but $\mathcal{K}^* \neq -\mathcal{K}$. Finally, in the third case let us take $\langle u, \tilde{u} \rangle_{H_0^2} := \int_\Omega \Delta u \Delta \tilde{u} \, dx$ and $\mathcal{K} = \{v \in H_0^2(\Omega) : v \geq 0\}$. Then $\mathcal{K}^* = \{w \in H_0^2(\Omega) : (-\Delta)^2 w \leq 0 \text{ weakly}\}$. In this case it is in general not true that $\mathcal{K}^*$ is a subset of $-\mathcal{K}$. It is however true for the case where $\Omega = B_1(0)$ as the next result shows.

Lemma 4.13 *Let* $\Omega = B_1(0)$ *and consider the Hilbert space* $H = H_0^m(B_1(0))$ *with* $\langle u, \tilde{u} \rangle := \int_{B_1(0)} \mathcal{D}^m u \mathcal{D}^m \tilde{u} \, dx$. *Let* $\mathcal{K} = \{v \in H : v \geq 0\}$. *Then* $\mathcal{K}^* \subset -\mathcal{K}$ *and, more precisely, for every* $w \in \mathcal{K}^*$ *we have* $w < 0$ *or* $w \equiv 0$ *in* $B_1(0))$.

Proof We only prove the statement $\mathcal{K}^* \subset -\mathcal{K}$. The more precise formulation $w < 0$ or $w \equiv 0$ for all $w \in \mathcal{K}^*$ can be found in [13]. Let $w \in \mathcal{K}^*$ and let $\phi \in L^2(B_1(0))$ with $\phi \geq 0$ and consider the solution $v_\phi \in H_0^m(B_1(0))$ of

$$(-\Delta)^m v_\phi = \phi \text{ in } B_1(0), \quad v_\phi = \frac{\partial}{\partial \nu} v_\phi = \ldots = \left(\frac{\partial}{\partial \nu}\right)^{m-1} v_\phi = 0 \text{ on } \partial B_1(0).$$

By Lemma 4.10, $v_\phi \in \mathcal{K}$, so that $0 \geq \langle w, v_\phi \rangle = \int_{B_1(0)} \phi w \, dx$. Since this holds for all $\phi \in L^2(B_1(0))$ with $\phi \geq 0$ we conclude $w \leq 0$. □

Finally, we are ready to prove simplicity, positivity and symmetry of first eigenfunctions.

Theorem 4.14 *The first eigenvalue* λ_1 *of* $(-\Delta)^m$, $m \geq 1$ *on* $B_1(0)$ *with Dirichlet boundary conditions is simple. Any corresponding eigenfunction is of one sign and radially symmetric.*

Proof Let $u \in H_0^m(B_1(0))$ be an eigenfunction corresponding to λ_1. Assume, for contradiction, that u is sign-changing. We use the scalar product $\langle u, \tilde{u} \rangle := \int_{B_1(0)} \mathcal{D}^m u \mathcal{D}^m \tilde{u} \, dx$ and split $u = u_1 + u_2$ according to Theorem 4.12 with $\mathcal{K} = \{v \in H_0^m(B_1(0)) : v \geq 0\}$. The assumption that u is sign-changing implies $u_1 \in \mathcal{K} \setminus \{0\}$ and $u_2 \in \mathcal{K}^* \setminus \{0\}$. Lemma 4.13 implies that $u_2 < 0$ in $B_1(0)$. Therefore, if we set $\tilde{u} := u_1 - u_2$, then $\tilde{u} \in \mathcal{K}$ and moreover

$$u_2 - u_1 \leq u_2 \leq u_1 + u_2 < u_1 - u_2 \text{ a.e. in } B_1(0).$$

Moreover, the first inequality is strict on a set of positive measure. Therefore $\|u_1 + u_2\|_{L^2} < \|u_1 - u_2\|_{L^2}$. However, due to $u_1 \perp_{H_0^m} u_2$ we see that $\|u_1 + u_2\|_{H_0^m}^2 =$

$\|u_1 - u_2\|^2_{H^m_0}$. This contradicts the minimality of λ_1. Therefore, any eigenfunction corresponding to λ_1 does not change sign and in particular it is either strictly positive ($u \geq 0$ solving the eigenvalue equation implies $-u \in \mathcal{K}^*$) or strictly negative (nontrivial elements of $\mathcal{K}^*$ are strictly negative).

A standard argument allows now to deduce simplicity from positivity. Suppose that $u, \tilde{u}$ are linearly independent eigenfunctions corresponding to λ_1. Then w.l.o.g $u, \tilde{u}$ are both strictly positive. However, $u - \alpha\tilde{u}$ for $\alpha \in \mathbb{R}$ is also an eigenfunction for λ_1, and it is sign-changing for suitable $\alpha > 0$. This contradiction shows simplicity.

Finally, radial symmetry of the eigenfunction is a consequence of simplicity. $\quad\square$

4.3 Symmetry for Nonlinear Problems by Uniqueness and Non-resonance

In this section we prove two statements that directly imply symmetry for solutions of the nonlinear boundary value problem

$$(-\Delta)^m u = f(x, u) \text{ in } B_1(0), \quad u = \frac{\partial}{\partial\nu}u = \ldots = \left(\frac{\partial}{\partial\nu}\right)^{m-1} u = 0 \text{ on } \partial B_1(0). \tag{4.4}$$

Recall the eigenvalue sequence $\lambda_1 < \lambda_2 \leq \lambda_3 \leq \ldots$ for the polyharmonic operator $(-\Delta)^m$ on the unit ball with Dirichlet boundary conditions.

Theorem 4.15 (Uniqueness) *Let $f : B_1(0) \times \mathbb{R} \to \mathbb{R}$ be a Carathéodory-function with the following properties*

(1) $|x| = |y| \Rightarrow f(x, s) = f(y, s)$ for all $s \in \mathbb{R}$ and all $x, y \in B_1(0)$,
(2) $f(\cdot, 0) \in L^2(B_1(0))$,
(3) Either $\exists L, \epsilon > 0$ such that

$$(*)_1 \quad -L \leq \frac{f(x, s) - f(x, t)}{s - t} \leq \lambda_1 - \epsilon \quad \text{for all } s, t \in \mathbb{R}, s \neq t, x \in B_1(0)$$

or $\exists L, \epsilon > 0, i \geq 2$ such that
$$(*)_i$$
$$\lambda_{i-1} + \epsilon \leq \frac{f(x, s) - f(x, t)}{s - t} \leq \lambda_i - \epsilon \quad \text{for all } s, t \in \mathbb{R}, s \neq t, x \in B_1(0).$$

Then (4.4) has a unique weak solution $u \in H^m_0(B_1(0))$. This solution is necessarily radially symmetric.

Proof We give the proof only in case of $(*)_i$ for $i \geq 2$. The case $(*)_1$ differs only in a few details. Clearly $\lambda_{i-1} < \lambda_i$. The inequality

$$|f(x, s)| \leq |f(x, 0)| + |s|\lambda_i$$

and assumption (2) guarantee that $u \in L^2(B_1(0))$ implies $f(\cdot, u) \in L^2(B_1(0))$. Let $\kappa := \frac{\lambda_{i-1} + \lambda_i}{2}$ and set $g(x, t) := f(x, t) - \kappa t$. Then assumption (3) implies

$$\frac{\lambda_{i-1} - \lambda_i + 2\epsilon}{2} \leq \frac{g(x, s) - g(x, t)}{s - t} \leq \frac{\lambda_i - \lambda_{i-1} - 2\epsilon}{2} \tag{4.5}$$

for all $s, t \in \mathbb{R}, s \neq t, x \in B_1(0)$. Let $T_\kappa : L^2(B_1(0)) \to H_0^m(B_1(0))$ be the solution operator mapping the right-hand side $h \in L^2(B_1(0))$ to the solution $w \in H_0^m(B_1(0))$ of the problem

$$(-\Delta)^m w - \kappa w = h \text{ in } B_1(0), \quad w = \frac{\partial}{\partial \nu} w = \ldots = \left(\frac{\partial}{\partial \nu}\right)^{m-1} w = 0 \text{ on } \partial B_1(0).$$

Notice that T_κ is well-defined and continuous since κ is not a Dirichlet eigenvalue of $(-\Delta)^m$. The map $T_\kappa : L^2(B_1(0)) \to L^2(B_1(0))$ is compact and its operator norm can be computed as

$$\|T_\kappa\| = \sup_{h \neq 0} \frac{\|T_\kappa h\|_{L^2}}{\|h\|_{L^2}} = \frac{1}{\min\{|\lambda_j - \kappa| : j \in \mathbb{N}\}} = \frac{2}{\lambda_i - \lambda_{i-1}}.$$

Next, consider the operator

$$G := \begin{cases} L^2(B_1(0)) \to L^2(B_1(0)), \\ \quad u \mapsto g(\cdot, u). \end{cases}$$

Then $u \in H_0^m(B_1(0))$ is a weak solution of (4.4) if and only if $u = T_\kappa(G(u))$, $u \in L^2(B_1(0))$. Hence, it is enough to show that $T_\kappa \circ G$ is a contraction in $L^2(B_1(0))$, i.e.,

$$\|T_\kappa(G(u)) - T_\kappa(G(v))\|_{L^2} \leq \|T_\kappa\| \, \|G(u) - G(v)\|_{L^2}$$

$$\overset{(4.5)}{\leq} \underbrace{\frac{2}{\lambda_i - \lambda_{i-1}} \frac{\lambda_i - \lambda_{i-1} - 2\epsilon}{2}}_{<1} \|u - v\|_{L^2}.$$

By the contraction mapping principle this establishes existence and uniqueness. Notice that if $u(\cdot)$ solves (4.4) then by assumption (1) the function $u(A\cdot)$ also solves (4.4) for every orthogonal matrix $A \in O(n)$. By uniqueness we get $u(x) = u(Ax)$, which establishes radial symmetry. $\qquad\square$

Infinitely many eigenvalues of the above sequence have radially symmetric eigenfunctions. But there is also a sequence $\{\nu_i\}_{i \in \mathbb{N}}$ of eigenvalues corresponding to non-radially symmetric eigenfunctions. The idea behind the next result (Theorem 4.20) goes back to [18] and can be explained as follows: if the nonlinearity

f in (4.4) is non-resonant w.r.t. the non-radial Dirichlet spectrum of $(-\Delta)^m$ on Ω then all solutions of (4.4) must be radially symmetric. A precise meaning of *non-resonant* will be given below in Theorem 4.20.

Definition 4.16 Let us denote by $L^2_{rad}(B_1(0))$ the space of all radially symmetric functions in $L^2(B_1(0))$, and let $Y = L^2_{rad}(B_1(0))^\perp$ be its L^2-orthogonal complement. The operator $(-\Delta)^m : H^{2m}(B_1(0)) \cap H^m_0(B_1(0)) \cap L^2_{rad}(B_1(0)) \to L^2_{rad}(B_1(0))$ is selfadjoint and has (simple) eigenvalues $\{\lambda^{rad}_i\}_{i\in\mathbb{N}}$ with corresponding radially symmetric eigenfunctions $\{\phi^{rad}_i\}_{i\in\mathbb{N}}$ building an orthonormal basis of $L^2_{rad}(B_1(0))$. Let $\{\mu_i\}_{i\in\mathbb{N}}$ be the set of non-radial eigenvalues, i.e., eigenvalues for which there exists at least one eigenfunction belonging to Y. We denote by $\{\psi_i\}_{i\in\mathbb{N}}$ an ONB of Y consisting of non-radial eigenfunctions corresponding to $\{\mu_i\}_{i\in\mathbb{N}}$.

Remark 4.17

1. Note that $\lambda_1 = \lambda^{rad}_1$ by Theorem 4.14 and hence $\mu_1 \geq \lambda_2$.
2. $L^2(B_1(0)) = L^2_{rad}(B_1(0)) \oplus L^2_{rad}(B_1(0))^\perp$ with $L^2_{rad}(B_1(0)) = \text{span}\{\phi^{rad}_i : i \in \mathbb{N}\}$ and $L^2_{rad}(B_1(0))^\perp = \text{span}\{\psi_i : i \in \mathbb{N}\}$.
3. We do not exclude $\{\lambda^{rad}_i\}_{i\in\mathbb{N}} \cap \{\mu_i\}_{i\in\mathbb{N}} \neq \emptyset$. In the intersection could be an eigenvalue which has both radial and non-radial eigenfunctions. Its eigenspace has an ONB with elements from $\{\phi^{rad}_i\}_{i\in\mathbb{N}}$ as well as from $\{\psi_i\}_{i\in\mathbb{N}}$. According to the above definition such an eigenvalue is called *non-radial*.

Lemma 4.18 *Let $h \in Y = L^2_{rad}(B_1(0))^\perp$ and $\kappa \notin \{\mu_i\}_{i\in\mathbb{N}}$. Then, in the space $H^m_0(B_1(0)) \cap Y$ there exists a unique element $w =: T_\kappa h$ solving*

$$(-\Delta)^m w - \kappa w = h \text{ in } B_1(0), \quad w = \frac{\partial}{\partial\nu}w = \ldots = \left(\frac{\partial}{\partial\nu}\right)^{m-1}$$

$$w = 0 \text{ on } \partial B_1(0). \tag{4.6}$$

Moreover,

$$\|T_\kappa\| = \sup_{h\neq 0}\frac{\|T_\kappa h\|_{L^2}}{\|h\|_{L^2}} = \frac{1}{\min\{|\mu_j - \kappa| : j \in \mathbb{N}\}}.$$

Proof Every element $h \in Y$ can be expressed by the basis $\{\psi_i\}_{i\in\mathbb{N}}$, i.e., $h = \sum_{i\in\mathbb{N}}\langle h, \psi_i\rangle_{L^2}\psi_i$ and hence, by the non-resonance assumption on κ, we have that

$$w = \sum_{i\in\mathbb{N}}\frac{\langle h, \psi_i\rangle_{L^2}}{\mu_i - \kappa}\psi_i$$

solves (4.6). Hence

$$\|w\|^2_{L^2} = \sum_{i\in\mathbb{N}}\frac{\langle h, \psi_i\rangle^2_{L^2}}{(\mu_i - \kappa)^2} \leq \frac{1}{\min\{|\mu_j - \kappa|^2 : j \in \mathbb{N}\}}\|h\|^2_{L^2}$$

which shows that $\|T_\kappa\| \leq \frac{1}{\min\{|\mu_j - \kappa|: j \in \mathbb{N}\}}$. The fact that equality holds can be seen by choosing $h = \psi_{j_0}$ where $j_0 \in \mathbb{N}$ is the index where $|\mu_{j_0} - \kappa| = \min\{|\mu_j - \kappa| : j \in \mathbb{N}\}$. $\square$

Lemma 4.19 *Let $\phi \in H_0^m(B_1(0)) \cap L_{rad}^2(B_1(0))$ and $\psi \in H_0^m(B_1(0)) \cap Y$. Then*

$$\int_{B_1(0)} \mathscr{D}^m \phi \, \mathscr{D}^m \psi \, dx = 0.$$

Proof By density it is enough to show the claim when ϕ is a radial $C_0^\infty(B_1(0))$ function. In this case integration by parts shows that $\int_{B_1(0)} \mathscr{D}^m \phi \, \mathscr{D}^m \psi \, dx = \int_{B_1(0)} (-\Delta)^m \phi \psi \, dx$. The latter vanishes because $(-\Delta)^m \phi$ is radial and hence belongs to $L_{rad}^2(B_1(0))$ whereas ψ belongs to $Y = L_{rad}^2(B_1(0))^\perp$. $\square$

Theorem 4.20 (Nonresonance w.r.t. Non-radial Spectrum) *Let $f : [0, \infty) \times \mathbb{R} \to \mathbb{R}$ be a Carathéodory-function with the following properties*

(1) $|x| = |y| \Rightarrow f(x, s) = f(y, s)$ *for all $s \in \mathbb{R}$ and all $x, y \in B_1(0)$,*
(2) $f(\cdot, 0) \in L^2(B_1(0))$,
(3) Either $\exists L, \epsilon > 0$ such that

$$(**)_1 \quad -L \leq \frac{f(x, s) - f(x, t)}{s - t} \leq \mu_1 - \epsilon \quad \text{for all } s, t \in \mathbb{R}, s \neq t, x \in B_1(0)$$

or $\exists L, \epsilon > 0, i \geq 2$ such that
$(**)_i$

$$\mu_{i-1} + \epsilon \leq \frac{f(x, s) - f(x, t)}{s - t} \leq \mu_i - \epsilon \quad \text{for all } s, t \in \mathbb{R}, s \neq t, x \in B_1(0).$$

Then every weak solution of (4.4) is radially symmetric.

Proof As in the proof of Theorem 4.15 we only deal with the case of $(**)_i$ with $i \geq 2$. Consider orthogonal projections $P : L^2(B_1(0)) \to L_{rad}^2(B_1(0))$ and $Q : L^2(B_1(0)) \to L_{rad}^2(B_1(0))^\perp$. Let $\kappa = \frac{\mu_{i-1} + \mu_i}{2}$ and set $g(x, t) := f(x, t) - \kappa t$. If we split $u \in H_0^m(B_1(0))$ into $u = v + w$ with $v \in H_0^m(B_1(0)) \cap L_{rad}^2(B_1(0))$ and $w \in H_0^m(B_1(0)) \cap Y$ then (4.4) can be split into two equations

$$(-\Delta)^m v - \kappa v = Pg(x, v + w) \text{ in } B_1(0), \quad v = \ldots = \left(\frac{\partial}{\partial \nu}\right)^{m-1}$$

$$v = 0 \text{ on } \partial B_1(0), \tag{4.7}$$

$$(-\Delta)^m w - \kappa w = Qg(x, v + w) \text{ in } B_1(0), \quad w = \ldots = \left(\frac{\partial}{\partial \nu}\right)^{m-1}$$

$$w = 0 \text{ on } \partial B_1(0). \tag{4.8}$$

In fact, (4.7) and (4.8) follow by Lemma 4.19 from the weak form of (4.4) and they are understood as

$$\int_{B_1(0)} \left(\mathscr{D}^m v \, \mathscr{D}^m \phi - \kappa v \phi \right) dx = \int_{\Omega} g(x, v + w) \phi \, dx \quad \text{for all}$$

$$\phi \in H_0^m(B_1(0))) \cap L^2_{rad}(B_1(0))$$

and

$$\int_{B_1(0)} \left(\mathscr{D}^m w \, \mathscr{D}^m \psi - \kappa w \psi \right) dx = \int_{\Omega} g(x, v + w) \psi \, dx \quad \text{for all}$$

$$\psi \in H_0^m(B_1(0))) \cap L^2_{rad}(B_1(0))^{\perp},$$

respectively. Using the solution operator $T_\kappa : Y \to Y$ from Lemma 4.18 and the operator $G : L^2(B_1(0)) \to L^2(B_1(0))$ as defined in the proof of Theorem 4.15 we see that (4.8) is equivalent to

$$w = (T_\kappa \circ Q \circ G)(v + w). \tag{4.9}$$

For every fixed $v \in H_0^m(B_1(0)) \cap L^2_{rad}(B_1(0))$, we can show exactly like in Theorem 4.15 that the map $w \mapsto (T_\kappa \circ Q \circ G)(v + w)$ is a contraction on Y. Hence (4.9) has a unique solution. Notice that $w = 0$ solves (4.9) since $G(v) = f(v) - \kappa v \in L^2_{rad}(B_1(0))$ and so $Q(G(v)) = 0$. Thus, all solutions $u = v + w$ of (4.4) necessarily have $w = 0$ and are therefore radial. $\qquad \square$

4.4 Symmetry via Moving Plane Method: An Example from Potential Theory

Let us step back more than three centuries into the year 1686/1687 when Isaac Newton published the first edition of *Principia Mathematica*. In this book he developed among many other seminal ideas the concept of the gravitational potential u of a body $\Omega \subset \mathbb{R}^3$ with mass-density $\rho : \Omega \to [0, \infty)$. In nowadays mathematical language we write it in the form

$$u(x) = \frac{1}{4\pi} \int_{\Omega} \frac{\rho(y)}{|x - y|} dy, \quad x \in \mathbb{R}^3.$$

If Ω is a ball $B_R(0)$ and if the mass density is a constant $\rho > 0$ then Newton computed

$$u(x) = \begin{cases} \rho\left(\dfrac{R^2}{2} - \dfrac{|x|^2}{6}\right), & |x| \leq R, \\ \dfrac{\rho R^3}{3|x|}, & |x| \geq R. \end{cases}$$

Two observations are important:

1. Outside the ball the gravitational potential coincides with that of a single point centered at the origin whose mass equals the mass of the entire ball.
2. The potential u is radially symmetric. In particular: $u|_{\partial B_R(0)}$ is constant.

The first observation (and its generalization to radially symmetric mass densities) allows to reduce celestial mechanics of stars and planets to the interaction of point masses. The second observation has an impact that one can check: the gravitational force $F = -\nabla u$ exerted upon a unit point mass by the body $B_R(0)$ is always normal to the surface and has everywhere on the surface the same value. In other words: wherever you travel on the surface of a ball-shaped planet and carry a unit mass with you it always weighs the same and when you let it drop then it falls perpendicularly to the surface of the planet (the surface being a level set of the gravitational potential). The second observation also triggers the following mathematical question.

> Are there bodies $\Omega \subset \mathbb{R}^3$ different from the ball with the property that the constant-density gravitational potential is constant everywhere on $\partial \Omega$?

Surprisingly, the question was answered mathematically rigorously only in 2000—more than 300 years after Newton. In fact it was answered for the generalization of the Newtonian potential to n dimensions.

Theorem 4.21 (Fraenkel [12]) *Let $\Omega \subset \mathbb{R}^n$ be a bounded open set and let ω_n be the surface measure of the unit-sphere in $\mathbb{R}^n$. Consider*

$$u(x) = \begin{cases} \dfrac{1}{2\pi} \displaystyle\int_\Omega \log \dfrac{1}{|x - y|}\, dy, & n = 2, \\ \dfrac{1}{(n-2)\omega_n} \displaystyle\int_\Omega \dfrac{1}{|x - y|^{n-2}}\, dy, & n \geq 3. \end{cases}$$

If u is constant on $\partial \Omega$ then Ω is a ball.

The goal of this chapter is to prove yet another generalization of Fraenkel's theorem, cf. [22]. Note that no boundary-regularity of Ω is needed in Fraenkel's theorem. This is quite remarkable and distinguishes this result from many other results on overdetermined problems. Unlike in Fraenkel's theorem for our result we need to a-priori restrict the class of open sets. It is open how to overcome the restriction to convex bounded domains, cf. Remark 4.23 below for a partial answer to this question.

Theorem 4.22 *Let $\Omega \subset \mathbb{R}^n$ be a bounded convex domain. For $\alpha \geq 2$ consider*

$$u(x) = \begin{cases} \displaystyle\int_\Omega \log \frac{1}{|x-y|} \, dy, & n = \alpha, \\[3mm] \displaystyle\int_\Omega \frac{1}{|x-y|^{n-\alpha}} \, dy, & n \neq \alpha. \end{cases} \tag{4.10}$$

If u is constant on $\partial\Omega$ then Ω is a ball.

Remark 4.23

1. The potential $\int_\Omega \frac{1}{|x-y|^{n-\alpha}} \, dy$ is called a Riesz-potential with unit density.
2. Lu and Zhu [20] have found another version of Theorem 4.22, where instead of the convexity of Ω they assumed $\partial\Omega \in C^1$. Moreover, in the follow-up paper [15] the analysis was extended to the case of Bessel potentials.

It is easy to see that the converse of both Theorems 4.21 and 4.22 hold: suppose $\Omega = B_R(0) \subset \mathbb{R}^n$ is a ball centered at the origin. Then u is radially symmetric and hence u is constant on $\partial\Omega$.

Let us begin with the crucial observation that $\partial\Omega = u^{-1}\{\beta\}$ where $\beta := u|_{\partial\Omega}$.

Lemma 4.24 *Let u, Ω and $\alpha \geq 2$ be as in Theorem 4.22 and $\beta := u|_{\partial\Omega}$. If $n \geq \alpha$ then $u > \beta$ in Ω and $u < \beta$ in $\overline{\Omega}^C$. If $n < \alpha$ then $u < \beta$ in Ω and $u > \beta$ in $\overline{\Omega}^C$.*

Remark 4.25 In Fraenkel's case $\alpha = 2$ this lemma was proven under the sole assumption Ω open, bounded. In the case $\alpha > 2$ we use convexity of Ω to establish this result.

Proof We begin with the remark that Riesz-potentials are $C^l(\mathbb{R}^n)$ for all $0 < l < \alpha$. We assume w.l.o.g that $\alpha > 2$ ($\alpha = 2$ corresponds to Fraenkel's theorem) so that we have $u \in C^2(\mathbb{R}^n)$. Now let us restrict to the case $2 < \alpha \leq n$. The remaining case $\alpha > n$ works in a very similar way. Note that $\Delta|x|^{\alpha-n} = (\alpha-n)(\alpha-2)|x|^{\alpha-n-2} < 0$ and $\Delta \log \frac{1}{|x|} = (2-n)|x|^{-2} < 0$. Therefore u is superharmonic on $\mathbb{R}^n$ and hence inside Ω the function u is larger than the value β of u on $\partial\Omega$. It remains to consider u outside Ω. We show that the convexity of Ω implies that u has no local extremum outside Ω. Since either $u(x) \to 0$ (for $\alpha < n$) or $u(x) \to -\infty$ (for $\alpha = n$) as $|x| \to \infty$ the absence of local extrema implies that u is smaller than β outside Ω and we are done. So let $x \in \mathbb{R}^n \setminus \overline{\Omega}$. By the convexity of Ω we can separate x from Ω through a hyperplane, i.e., there exists a unit vector $e \in \mathbb{R}^n$ and a point $z_0 \in \mathbb{R}^n \setminus \overline{\Omega}$ such that

$$(y - z_0) \cdot e < 0 < (x - z_0) \cdot e \quad \text{for all } y \in \Omega.$$

In particular $(x - y) \cdot e > 0$ for all $y \in \Omega$. Since

$$\nabla u(x) \cdot e = c_{\alpha,n} \int_\Omega \frac{(x - y) \cdot e}{|x - y|^{n-\alpha+1}} \, dy$$

and the integrand is strictly positive we see that u has no local extremum outside Ω.

□

As a consequence, one can rewrite the definition of u as follows. Let $f_H : \mathbb{R} \to \mathbb{R}$ be the Heaviside-function, i.e., $f_H(t) := 1$ for $t > 0$ and $f_H(t) := 0$ for $t \leq 0$. Thus, Lemma 4.24 can be phrased as

$$\chi_\Omega = \begin{cases} f_H(u - \beta), & n \geq \alpha, \\ f_H(\beta - u), & n < \alpha. \end{cases}$$

And now one can rewrite the potential u as

$$u(x) = \begin{cases} \displaystyle\int_{\mathbb{R}^n} \log \frac{1}{|x - y|} f_H(u(y) - \beta) \, dy, & n = \alpha, \\ \displaystyle\int_{\mathbb{R}^n} \frac{f_H(u(y) - \beta)}{|x - y|^{n-\alpha}} \, dy, & n > \alpha, \\ \displaystyle\int_{\mathbb{R}^n} \frac{f_H(\beta - u(y))}{|x - y|^{n-\alpha}} \, dy, & n < \alpha, \end{cases} \qquad (4.11)$$

i.e., one can consider u to be the solution of the **nonlinear** integral equation (4.11). Notice that the unknown domain Ω has disappeared from (4.11). The main idea of the proof is to show radial symmetry of the solution u of (4.11) via the method of moving planes—suitably adapted for nonlinear integral equations.

Remark 4.26

1. The original moving plane method goes back to Alexandrov [2] and Serrin [23] and was further developed by Gidas, Ni, Nirenberg [14] for second order equations.
2. Recent improvements of the moving plane method for higher order equations and pseudo differential operators using integral representations rather than local maximum principles were achieved by Chang, Yang [9], Berchio, Gazzola, Weth [4], Li [19], Chen, Li, Ou [10] and Birkner, López-Mimbela, Wakolbinger [5].

Before we prove the main result let us make some connections to polyharmonic boundary value problems.

Definition 4.27 (Polyharmonic Green Function in $\mathbb{R}^n$) Let $x, y \in \mathbb{R}^n$. Define

$$G(x, y) := k_{m,n} |x - y|^{2m-n}, \qquad k_{m,n} = \frac{\Gamma(\frac{n}{2} - m)}{2^{2m} \pi^{n/2} \Gamma(m)}$$

if $n > 2m$ or $n < 2m$ and n odd

and

$$G(x, y) := k_{m,n}|x - y|^{2m-n} \log \frac{1}{|x-y|}, \quad k_{m,n} = \frac{(-1)^{m-\frac{n}{2}}}{2^{2m-1}\pi^{n/2}\Gamma(m)}$$

if $n \leq 2m$ and n even.

Then G is the Green function of $(-\Delta)^m$ on $\mathbb{R}^n$.

Remark 4.28 Let Ω be open, bounded and $U(x) = \int_\Omega G(x, y)\,dy$ for $x \in \mathbb{R}^n$. Then $U \in C^{2m}(\Omega)$ with $(-\Delta)^m U = 1$ on Ω and $U \in C^{2m}(\mathbb{R}^n \setminus \overline{\Omega})$ and $(-\Delta U) = 0$ on $\mathbb{R}^n \setminus \overline{\Omega}$. Moreover $U \in C^l(\mathbb{R}^n)$ for $0 < l < 2m$ and $(-\Delta)^m U = \chi_\Omega$ in $\mathbb{R}^n$ in the distributional sense.

Hence, if we assume that $\alpha = 2m$ then in the case $n < 2m$, n odd, or $n \geq 2m$ the potential u from Theorem 4.22 weakly satisfies

$$(-\Delta)^m u = \frac{1}{k_{m,n}}\chi_\Omega \text{ in } \mathbb{R}^n.$$

If we combine this with Lemma 4.24 then the above equation can be restated as

$$(-\Delta)^m u = \frac{1}{k_{m,n}} \begin{cases} f_H(u - \beta) \text{ in } \mathbb{R}^n, & n \geq 2m, \\ f_H(\beta - u) \text{ in } \mathbb{R}^n, & n < 2m, \quad n \text{ odd}, \end{cases} \tag{4.12}$$

i.e., u (weakly) satisfies the **semilinear** polyharmonic equation (4.12) on $\mathbb{R}^n$ where the reference to the unknown domain Ω has disappeared. Therefore, Theorem 4.22 can be seen as an example, where radial symmetry for positive solutions of semilinear polyharmonic equations can be shown by the moving plane method. However, as the moving plane method for second order elliptic differential equations is based on the maximum principle which is unavailable for higher order equations, the way to proceed for (4.12) is to adapt the moving plane method for the nonlinear integral equation (4.11).

After this comment we can now begin with the moving-plane proof of Theorem 4.22.

4.4.1 Asymptotic Expansion and the Role of the Barycenter

Let us define the barycenter of Ω by

$$B(\Omega) := \frac{1}{\text{vol }\Omega} \int_\Omega y\,dy.$$

Lemma 4.29 *Let u, Ω be as in Theorem 4.22. Then*

$$u(x) = \begin{cases} \operatorname{vol} \Omega \left(\log \dfrac{1}{|x|} - |x|^{-2}x \cdot B(\Omega) \right) + h(x) & \text{if } n = \alpha, \\[3mm] \operatorname{vol} \Omega \left(|x|^{\alpha-n} + (n-\alpha)|x|^{\alpha-n-2}x \cdot B(\Omega) \right) + h(x) & \text{if } n \neq \alpha \end{cases}$$

$$(4.13)$$

where h satisfies $|h(x)| \leq C|x|^{\alpha-n-2}$, $|\nabla h(x)| \leq C|x|^{\alpha-n-3}$ for some constant $C > 0$.

Proof Let $n \neq \alpha$. A direct application of Taylor's theorem to $g(t) := |x - ty|^{\alpha-n}$ yields

$$|x - y|^{\alpha-n} = |x|^{\alpha-n} - (\alpha - n)|x|^{\alpha-n-2}x \cdot y + k(x, y) \qquad (4.14)$$

where there exists a constant $C > 0$ and a radius $R_0 > 0$ such that

$$|k(x, y)| \leq C|x|^{\alpha-n-2}, \quad |\nabla_x k(x, y)| \leq C|x|^{\alpha-n-3} \qquad \text{for all } |x| \geq R_0, y \in \Omega. \qquad (4.15)$$

Here $R_0 > 0$ is chosen such that $\overline{\Omega} \subset B_{R_0}(0)$. The claim of the lemma follows from integrating (4.14). The estimate for $h(x) := \int_\Omega k(x, y)\, dy$ follows from (4.15). The proof for $n = \alpha$ is similar.   □

The lemma has the following consequence: if u, Ω are like in Theorem 4.22 then for arbitrary $q \in \mathbb{R}^n$ the function $u(\cdot + q)$ and the domain $\Omega - q$ also satisfy the assumption of Theorem 4.22. Hence, w.l.o.g, we may assume that $B(\Omega) = 0$ and, as a consequence, that the middle term in the asymptotic expansion (4.13) vanishes.

4.4.2 The Moving Plane Method

For a point $x = (x_1, x') \in \mathbb{R}^n$ let $x^\lambda = (2\lambda - x_1, x')$ be the reflection of x at the hyperplane $T_\lambda := \{x \in \mathbb{R}^n : x_1 = \lambda\}$. Hence $|x^\lambda|^2 - |x|^2 = 4\lambda(\lambda - x_1)$. Also define the half-space $H_\lambda := \{x \in \mathbb{R}^n : x_1 < \lambda\}$ and note that $\partial H_\lambda = T_\lambda$. Recall that u from Theorem 4.22 solves the nonlinear integral equation (4.11). Depending on the value α from Theorem 4.22 define on $\overline{H}_\lambda$ the function

$$w_\lambda(x) := u(x) - u(x^\lambda) \text{ if } \alpha \leq n \quad \text{and} \quad w_\lambda(x) := -u(x) + u(x^\lambda) \text{ if } \alpha > n.$$

We will show that the function w_λ satisfies

$$w_\lambda(x) > 0 \text{ in } H_\lambda, \qquad \frac{\partial w_\lambda}{\partial x_1}(x) < 0 \text{ on } T_\lambda \qquad (4.16)$$

for all $\lambda > 0$. By continuity this implies $u(x_1, x') \geq u(-x_1, x')$ for all $x \in \mathbb{R}^n$, $x_1 \leq 0$. The reverse inequalities also hold by repeating the moving plane argument with the $-x_1$-direction. Hence $u(-x_1, x') = u(x_1, x')$ for all $x \in \mathbb{R}^n$ and moreover u is strictly monotone in the positive x_1-direction. Repeating the moving-plane argument with an arbitrary unit-direction instead of the x_1-direction one obtains that the function u is radially symmetric with respect to the origin and moreover radially strictly monotone. Together with the fact that $\partial \Omega$ is a level-surface of the function u this implies that Ω must be a ball centered at the origin. Thus, Theorem 4.22 is proved if we show (4.16) for all values of $\lambda > 0$.

Let us now outline how the prototypical moving-plane procedure is used in order to prove (4.16) for all values of $\lambda > 0$. It is done in 5 steps:

1. for every $\lambda > 0$ one has $w_\lambda(x) > 0$ if $|x|$ is sufficiently large, cf. Lemma 4.30
2. $w_\lambda > 0$ in H_λ if λ is sufficiently large, cf. Lemma 4.31
3. $w_\lambda \geq 0$ and $\lambda > 0$ implies $w_\lambda > 0$, cf. Lemma 4.32
4. the set $J := \{\lambda > 0 : w_\lambda > 0 \text{ in } H_\lambda\}$ is open, cf. Lemma 4.33
5. $J = (0, \infty)$, cf. Lemma 4.34

We state the precise form of the lemmas next and directly give their proofs.

Lemma 4.30 *For every $\lambda > 0$ there exists a value $R(\lambda) > 0$ such that for all $x \in H_\lambda$ with $|x| \geq R(\lambda)$ we have $w_\lambda(x) > 0$. The function $R(\lambda)$ can be chosen such that $R(\lambda)$ is continuous and non-increasing in λ.*

Proof According to the value of α the proof needs to be divided into several cases. Here we only do the case $2 \leq \alpha < n$. Recall that $|x| < |x^\lambda|$ for $x \in H_\lambda$ and $\lambda > 0$. With h being the function from Lemma 4.29 we have

$$u(x) - u(x^\lambda) = \text{vol}\,\Omega\,(|x|^{\alpha-n} - |x^\lambda|^{\alpha-n}) + h(x) - h(x^\lambda)$$

Assume first that $|x^\lambda|^2 \leq 2|x|^2$. By convexity of the function $s \mapsto s^{\frac{\alpha-n}{2}}$ for $s > 0$ we have

$$|x|^{\alpha-n} - |x^\lambda|^{\alpha-n} > \frac{n-\alpha}{2}|x^\lambda|^{\alpha-n-2}4\lambda(\lambda - x_1) \geq C_1|x|^{\alpha-n-2}\lambda(\lambda - x_1)$$

where $C_1 := (n - \alpha)2^{\frac{\alpha-n}{2}}$. By Lemma 4.29 we have $|h(x) - h(x^\lambda)| \leq 2C|x|^{\alpha-n-3}(\lambda - x_1)$. Hence

$$u(x) - u(x^\lambda) > |x|^{\alpha-n-3}(\lambda - x_1)\big(\text{vol}\,\Omega C_1|x|\lambda - 2C\big). \tag{4.17}$$

Next assume that $|x^\lambda|^2 \geq 2|x|^2$. Then

$$|x|^{\alpha-n} - |x^\lambda|^{\alpha-n} \geq |x|^{\alpha-n}(1 - 2^{\frac{\alpha-n}{2}}) =: C_2|x|^{\alpha-n},$$

where $C_2 > 0$. Again by Lemma 4.29 $|h(x) - h(x^\lambda)| \leq 2C|x|^{\alpha-n-2}$. Thus

$$u(x) - u(x^\lambda) \geq |x|^{\alpha-n}\big(\text{vol}\,\Omega C_2 - \frac{2C}{|x|^2}\big). \tag{4.18}$$

Hence, the right-hand sides both in (4.17) and (4.18) are positive provided

$$|x| > R(\lambda) := \max \left\{ \frac{2C}{\text{vol}\,\Omega C_1 \lambda}, \sqrt{\frac{2C}{\text{vol}\,\Omega C_2}} \right\}.$$

□

Lemma 4.31 *There exists $\lambda^* > 0$ such that for all $\lambda > \lambda^*$ we have $w_\lambda(x) > 0$ in H_λ.*

Proof We give the proof only in the case $2 \le \alpha < n$. Let $R(\lambda)$ be the function defined in Lemma 4.30. Let $c_1 := \min_{|x| \le R(1)} u(x)$. Hence $c_1 > 0$, and since $u(x)$ decays to 0 as $|x| \to \infty$ there exists a value $\lambda^* \ge 1$ such that $|x| \ge \lambda^*$ implies $u(x) \le c_1/2$. Let now $\lambda > \lambda^*$. Consider $x \in H_\lambda$ with $|x| > R(1)$. For such x we have $|x| > R(\lambda)$ and hence $u(x) > u(x^\lambda)$ by Lemma 4.30. Now consider $x \in H_\lambda$ with $|x| \le R(1)$. Since $|x^\lambda| \ge \lambda > \lambda^*$ we find $u(x) \ge c_1 > u(x^\lambda)$, and the claim is proved. □

The next lemma can be seen as an equivalent to the strong minimum principle.

Lemma 4.32 *Let $\lambda > 0$. If $w_\lambda \ge 0$ in H_λ then $w_\lambda > 0$ in H_λ and $\frac{\partial w_\lambda}{\partial x_1}(x) < 0$ on T_λ.*

Proof As usual we only treat $2 \le \alpha < n$. Let us first derive the two identities (4.20) and (4.21). By the nonlinear integral equation (4.11)

$$u(x) = \int_{\mathbb{R}^n} \frac{f_H(u(y) - \beta)}{|x-y|^{n-\alpha}}\,dy = \int_{H_\lambda} \dots dy + \int_{\mathbb{R}^n \setminus H_\lambda} \dots dy \qquad (4.19)$$

$$= \int_{H_\lambda} \frac{f_H(u(y) - \beta)}{|x-y|^{n-\alpha}} + \frac{f_H(u(y^\lambda) - \beta)}{|x-y^\lambda|^{n-\alpha}}\,dy.$$

Therefore

$$w_\lambda(x) = \int_{H_\lambda} \left(f_H(u(y) - \beta) - f_H(u(y^\lambda) - \beta) \right) \underbrace{\left(\frac{1}{|x-y|^{n-\alpha}} - \frac{1}{|x^\lambda - y|^{n-\alpha}} \right)}_{=:(*)}\,dy$$

$$(4.20)$$

and for $x \in T_\lambda$ we have $\frac{\partial w_\lambda}{\partial x_1}(x) = 2\frac{\partial u}{\partial x_1}(x)$ so that by (4.19)

$$\frac{1}{2}\frac{\partial w_\lambda}{\partial x_1}(x) = \int_{H_\lambda} f_H(u(y) - \beta) \underbrace{\frac{\partial}{\partial x_1}\left(\frac{1}{|x-y|^{n-\alpha}} \right)}_{=:(**)}$$

$$+ f_H(u(y^\lambda) - \beta)\frac{\partial}{\partial x_1}\left(\frac{1}{|x-y^\lambda|^{n-\alpha}} \right)\,dy. \qquad (4.21)$$

What do we need to conclude the statement of the lemma? Suppose we knew for all $x, y \in H_\lambda$ that $(*) > 0$ and for all $x \in T_\lambda$, $y \in H_\lambda$ that $(**) < 0$ as well as

$$\frac{\partial}{\partial x_1} \left(\frac{1}{|x - y|^{n-\alpha}} + \frac{1}{|x - y^\lambda|^{n-\alpha}} \right) = 0. \tag{4.22}$$

Then the statement of the lemma would follow like this: First, $w_\lambda \geq 0$ and $\neq 0$ by Lemma 4.30. Therefore $f_H(u(y) - \beta) - f_H(u(y^\lambda) - \beta)) \geq 0$ on H_λ and actually positive on a subset of H_λ of positive measure. Hence the integral in (4.20) is strictly positive for $x \in H_\lambda$. The same observation together with $(**) < 0$ allows to further estimate (4.21) for $x \in T_\lambda$

$$\frac{1}{2} \frac{\partial w_\lambda}{\partial x_1}(x) < \int_{H_\lambda} f_H(u(y^\lambda) - \beta) \frac{\partial}{\partial x_1} \left(\frac{1}{|x - y|^{n-\alpha}} + \frac{1}{|x - y^\lambda|^{n-\alpha}} \right) dy$$

$$= 0 \text{ by } (4.22)$$

and the claim would be shown. So what remains to be done?

Let us look at $(*)$: for $x, y \in H_\lambda$ one has

$$|x^\lambda - y|^2 = 4 \underbrace{(\lambda - x_1)}_{>0} \underbrace{(\lambda - y_1)}_{>0} + |x - y|^2$$

which implies $(*) > 0$ since $n > \alpha$.

Let us look at $(**)$: for $x \in T_\lambda$ and $y \in H_\lambda$ one has

$$\frac{\partial}{\partial x_1} |x - y| = \frac{x_1 - y_1}{|x - y|} = \frac{\lambda - y_1}{|x - y|} > 0$$

which implies $(**) > 0$ since $n > \alpha$.

Finally, let us look at (4.22): for fixed $y \in H_\lambda$, $x' \in \mathbb{R}^{n-1}$ the function $x_1 \mapsto \frac{1}{|x-y|^{n-\alpha}} + \frac{1}{|x-y^\lambda|^{n-\alpha}}$ is even around $x_1 = \lambda$. Therefore its derivative vanishes at $x_1 = \lambda$. $\qquad \square$

Lemma 4.33 *The set $J \subset (0, \infty)$ is open.*

Proof Assume that J is not open, i.e., there exists $\lambda_0 \in J$ which is an accumulation point of J^C. Thus, there exist sequences $\lambda_k \to \lambda_0$ as $k \to \infty$ and $x_k \in H_{\lambda_k}$ such that $w_{\lambda_k}(x_k) \leq 0$. Let $R(\lambda)$ be the function from Lemma 4.30. Clearly $|x_k| \leq R(\lambda_0/2)$, because $|x_k| > R(\lambda_0/2) \geq R(\lambda_k)$ for large k would imply $w_{\lambda_k}(x_k) > 0$ by Lemma 4.30, which cannot hold. Hence (by extracting a subsequence) we have $x_k \to x_0 \in \overline{H(\lambda_0)}$. Since $w_{\lambda_0} > 0$ in H_{λ_0} we must have $x_0 \in T_{\lambda_0}$. Thus, we find $2\frac{\partial u}{\partial x_1}(x_0) = \frac{\partial w_\lambda}{\partial x_1}(x_0) < 0$ due to Lemma 4.32. This contradicts $0 \geq w_\lambda(x_k) = u(x_k) - u(x_k^{\lambda_k})$ for large k. $\qquad \square$

Lemma 4.34 *The set $J = (0, \infty)$.*

Proof The set J is non-empty by Lemma 4.31. Let (μ, ∞) be the largest open interval contained in J. Then $w_\mu \geq 0$ in H_μ. Assume, for contradiction, that $\mu > 0$. By Lemma 4.32 we see that $w_\mu > 0$ in H_μ so that $\mu \in J$. A contradiction is reached since J is open by Lemma 4.33. ☐

4.5   An Example of Symmetry in an Overdetermined 4th Order Problem

The most famous second-order overdetermined problem was first studied by Serrin [23] and Weinberger [24]: let $\Omega \subset \mathbb{R}^n$ be a bounded C^2-domain and $u \in C^2(\overline{\Omega})$ a solution of

$$
\begin{aligned}
-\Delta u &= 1 \text{ in } \Omega, \\
u &= 0 \text{ on } \partial\Omega, \\
\frac{\partial u}{\partial \nu} &= \text{const.} < 0 \text{ on } \partial\Omega.
\end{aligned}
$$

Then Serrin, and independently Weinberger, proved that Ω is a ball and u is radially symmetric. Bennett [3] studied the following 4th order overdetermined problem:

$$
\begin{aligned}
\Delta^2 u &= -1 \text{ in } \Omega, \\
u &= \frac{\partial u}{\partial \nu} = 0 \text{ on } \partial\Omega, \\
\Delta u &= \text{const.} = c \text{ on } \partial\Omega.
\end{aligned}
\tag{4.23}
$$

Using Weinberger's method Bennett proved the following result.

Theorem 4.35 *Let Ω be a bounded $C^{4,\alpha}$-domain and suppose $u \in C^4(\overline{\Omega})$ solves (4.23). Then Ω is a ball of radius $|c|\sqrt{(n^2 + 2n)}$ and u is given by*

$$
\frac{-1}{2n}\left(\frac{(n+2)(nc)^2}{4} + \frac{nc}{2}r^2 + \frac{1}{4(n+2)}r^4 \right)
$$

where r denotes the distance of x to the center of the ball.

Recently, another proof of Serrin's result was given in [8] by Brandolini, Nitsch, Salani, Trombetti. Instead of giving Bennett's original proof, let us show in the case $c > 0$ how some of the ideas of [8] can be used to give an alternative proof of Bennett's result.

Remark 4.36

1. The alternative proof of Serrin's result given in [8] is also the basis for the answer to the stability question for Serrin's problem: if the normal derivative of u is close to a constant, how far is Ω from being a ball? A quantitative answer is given in [7]. It improves an earlier, more general result from [1] in the case of Serrin's problem.
2. Stability results for Bennett's problem seem to be unknown.

In [8] the authors are making use of the so-called Newton inequality, which can be stated as follows. Let A be a symmetric $n \times n$ matrix and $S_1(A) := \mathrm{trace}(A)$ and $S_2(A) := \frac{1}{2}(\mathrm{trace}(A)^2 - |A|^2) = \frac{1}{2}((\sum_j a_{jj})^2 - \sum_{i,j} a_{ij}^2)$ be the first two elementary symmetric functions of the eigenvalues of A. Then

$$S_2(A) \leq \frac{n-1}{2n} S_1(A)^2.$$

Equality holds if and only if A is a multiple of the identity matrix. This result and more on Newton's inequalities can be found in [16]. Let us denote by $S_2^{ij}(A) := \frac{\partial}{\partial a_{ij}} S_2(A) = \mathrm{trace}(A)\delta_{ij} - a_{ij}$. An important identity for $S_2(D^2\phi)$ and $\phi \in C^3(\Omega)$ is as follows

$$2S_2(D^2\phi) = \sum_{i=1}^{n} \frac{\partial}{\partial x_i} \left(\sum_{j=1}^{n} S_2^{ij}(D^2\phi) \frac{\partial \phi}{\partial x_j} \right)$$

$$= \mathrm{div}\left((\Delta\phi\, \mathrm{Id} - D^2\phi)\nabla\phi \right).$$

Therefore, Newton's inequality applied to $A = D^2\phi$ reads as follows

$$\mathrm{div}\left((\Delta\phi\, \mathrm{Id} - D^2\phi)\nabla\phi \right) = 2S_2(D^2\phi) \leq \frac{n-1}{n}(S_1(D^2\phi))^2$$

$$= \frac{n-1}{n}(\Delta\phi)^2. \tag{4.24}$$

Proof of Theorem 4.35 (Assuming $c > 0$:)

Step 1: $u < 0$ in Ω. Suppose, for contradiction, that somewhere in Ω the function u takes a nonnegative value. Then u attains its maximum at a point $q \in \Omega$. At this point we have $\underline{\Delta u}(q) \leq 0$. Since $\Delta u|_{\partial\Omega} = c > 0$ the function Δu attains its minimum over $\overline{\Omega}$ at a point $p \in \Omega$. Hence $\Delta^2 u(p) \geq 0$ in contradiction to (4.23).

Step 2: If u solves (4.23) then the following integral relation holds:

$$\int_{\Omega} u\, dx = -\int_{\Omega} (\Delta u)^2\, dx = \frac{-n}{n+4} c^2\, \mathrm{vol}(\Omega). \tag{4.25}$$

We start with a differential identity which holds for arbitrary functions $\phi, \psi \in C^3(\Omega)$:

$$\operatorname{div}\left(-\frac{1}{2}x\psi\,\Delta\phi - (\nabla\phi \cdot \nabla\psi)x + (\nabla\phi \cdot x)\nabla\psi + (\nabla\psi \cdot x)\nabla\phi\right)$$

$$= -\frac{n}{2}\psi\,\Delta\phi - \frac{1}{2}\psi\left(\nabla(\Delta\phi)\cdot x\right) + (2-n)\nabla\phi\cdot\nabla\psi + (\nabla\phi\cdot x)\Delta\psi + \frac{1}{2}(\nabla\psi\cdot x)\Delta\phi.$$

We choose $\phi = u$ and $\psi = \Delta u$, use the boundary condition $\nabla u = 0$, $\Delta u = c$ on $\partial\Omega$ and integrate to obtain

$$-\frac{n}{2}c^2\operatorname{vol}(\Omega) = \int_\Omega -\frac{n}{2}(\Delta u)^2 + (2-n)\nabla u \cdot \nabla(\Delta u) + (\nabla u \cdot x)\Delta^2 u\,dx$$

$$= \int_\Omega (\frac{n}{2} - 2)(\Delta u)^2 - (\nabla u \cdot x)\,dx$$

$$= \int_\Omega (\frac{n}{2} - 2)(\Delta u)^2 + nu\,dx$$

$$= -(\frac{n}{2} + 2)\int_\Omega (\Delta u)^2\,dx,$$

where for the last equality we have used the differential equation tested with u. This establishes the claim.

Step 3: If u solves (4.23) then the following integral relation holds:

$$\int_\Omega (\nabla u)^T D^2(\Delta u)\nabla(\Delta u)\,dx = -\frac{1}{4}\int_\Omega (\Delta u)^2 + \frac{c^2}{4}\operatorname{vol}(\Omega). \tag{4.26}$$

Since

$$\frac{1}{2}\operatorname{div}(\nabla u|\nabla\Delta u|^2) = \frac{1}{2}\Delta u|\nabla\Delta u|^2 + (\nabla u)^T D^2(\Delta u)\nabla(\Delta u)$$

$$= \frac{1}{4}\operatorname{div}\left((\Delta u)^2\nabla(\Delta u)\right) - \frac{1}{4}(\Delta u)^2\Delta^2 u + (\nabla u)^T D^2(\Delta u)\nabla(\Delta u)$$

the claim follows by integration using $\Delta^2 u = -1$ in Ω and $\nabla u = 0$, $\Delta u = c$ on $\partial\Omega$.

Step 4: If we apply Newton's inequality (4.24) to $\phi = \Delta u$ we obtain

$$\operatorname{div}\left((\Delta^2 u\,\operatorname{Id} - D^2(\Delta u))\nabla(\Delta u)\right) = 2S_2(D^2\Delta u) \leq \frac{n-1}{n}(\Delta^2 u)^2 = \frac{n-1}{n}.$$

Multiplying with u and using $u \leq 0$ we obtain after integration

$$\int_\Omega \nabla u \cdot \nabla(\Delta u) + (\nabla u)^T D^2(\Delta u)\nabla(\Delta u)\, dx \geq \frac{n-1}{n}\int_\Omega u\, dx.$$

One further integration by parts and (4.26) yields

$$-\frac{5}{4}\int_\Omega (\Delta u)^2 + \frac{c^2}{4}\operatorname{vol}(\Omega) \geq \frac{n-1}{n}\int_\Omega u\, dx.$$

But (4.25) tells us that in fact we have equality in the above inequality. Tracing this back to Newton's inequality we find that $D^2\Delta u$ is a multiple of the identity matrix, i.e., $D^2\Delta u = -\frac{1}{n}\delta_{ij}$ in Ω. If $a \in \Omega$ denotes the point where $A := \Delta u(a) = \max_\Omega \Delta u$ then

$$\Delta u(x) = \frac{1}{2n}(A - |x - a|^2).$$

If we now consider the boundary conditions for Δu then we see that Ω is a ball centered at a with radius $\sqrt{A - 2cn}$. The boundary condition for u and $\partial_\nu u$ then yield the stated form of u. $\qquad\square$

References

1. A. Aftalion, J. Busca, W. Reichel, Approximate radial symmetry for overdetermined boundary value problems. Adv. Differential Equ. **4**(6), 907–932 (1999)
2. A.D. Alexandrov, A characteristic property of spheres. Ann. Mat. Pura Appl. (4) **58**, 303–315 (1962)
3. A. Bennett, Symmetry in an overdetermined fourth order elliptic boundary value problem. SIAM J. Math. Anal. **17**(6), 1354–1358 (1986)
4. E. Berchio, F. Gazzola, T. Weth, Radial symmetry of positive solutions to nonlinear polyharmonic Dirichlet problems. J. Reine Angew. Math. **620**, 165–183 (2008)
5. M. Birkner, J.A. López-Mimbela, A. Wakolbinger, Comparison results and steady states for the Fujita equation with fractional Laplacian. Ann. Inst. H. Poincaré Anal. Non Linéaire **22**(1), 83–97 (2005)
6. T. Boggio, Sulle funzioni di green d'ordine m. Rend. Circ. Mat. Palerm **20**, 97–135 (1905)
7. B. Brandolini, C. Nitsch, P. Salani, C. Trombetti, On the stability of the Serrin problem. J. Differential Equ. **245**(6), 1566–1583 (2008)
8. B. Brandolini, C. Nitsch, P. Salani, C. Trombetti, Serrin-type overdetermined problems: an alternative proof. Arch. Ration. Mech. Anal. **190**(2), 267–280 (2008)
9. S.-Y.A. Chang, P.C. Yang, On uniqueness of solutions of nth order differential equations in conformal geometry. Math. Res. Lett. **4**(1), 91–102 (1997)
10. W. Chen, C. Li, B. Ou, Classification of solutions for an integral equation. Commun. Pure Appl. Math. **59**(3), 330–343 (2006)
11. L.C. Evans, Partial Differential Equations. Graduate Studies in Mathematics (American Mathematical Society, Providence, 1998), xviii+662 pp.

12. L.E. Fraenkel, *An Introduction to Maximum Principles and Symmetry in Elliptic Problems.* Cambridge Tracts in Mathematics, vol. 128 (Cambridge University Press, Cambridge, 2000)
13. F. Gazzola, H.-C. Grunau, G. Sweers, *Polyharmonic Boundary Value Problems.* Lecture Notes in Mathematics, vol. 1991 (Springer, Berlin, 2010). Positivity preserving and nonlinear higher order elliptic equations in bounded domains
14. B. Gidas, W.M. Ni, L. Nirenberg, Symmetry and related properties via the maximum principle. Commun. Math. Phys. **68**(3), 209–243 (1979)
15. X. Han, G. Lu, J. Zhu, Characterization of balls in terms of Bessel-potential integral equation. J. Differential Equ. **252**(2), 1589–1602 (2012)
16. G.H. Hardy, J.E. Littlewood, G. Pólya, *Inequalities.* Cambridge Mathematical Library (Cambridge University Press, Cambridge, 1988). Reprint of the 1952 edition
17. F. John, Partial Differential Equations. Applied Mathematical Sciences, 4th edn. (Springer, New York, 1982), x+249 pp.
18. A.C. Lazer, P.J. McKenna, A symmetry theorem and applications to nonlinear partial differential equations. J. Differential Equ. **72**(1), 95–106 (1988)
19. Y.Y. Li, Remark on some conformally invariant integral equations: the method of moving spheres. J. Eur. Math. Soc. **6**(2), 153–180 (2004)
20. G. Lu, J. Zhu, An overdetermined problem in Riesz-potential and fractional Laplacian. Nonlinear Anal. **75**(6), 3036–3048 (2012)
21. J.-J. Moreau, Décomposition orthogonale d'un espace hilbertien selon deux cônes mutuellement polaires. C. R. Acad. Sci. Paris **255**, 238–240 (1962)
22. W. Reichel, Characterization of balls by Riesz-potentials. Ann. Mat. Pura Appl. (4) **188**(2), 235–245 (2009)
23. J. Serrin, A symmetry problem in potential theory. Arch. Ration. Mech. Anal. **43**, 304–318 (1971)
24. H.F. Weinberger, Remark on the preceding paper of Serrin. Arch. Ration. Mech. Anal. **43**, 319–320 (1971)

Chapter 5
Recent Trends in Free Boundary Regularity

Henrik Shahgholian

Abstract In this lecture note I shall describe some recent developments of free boundary value problems and their regularity theory. Starting with the classical obstacle problem, and its development along with detailed proofs of each step of the theory, I shall go on to introduce various closely related semi-classical problems. The latter includes Bernoulli type FBP, Semi- and Quasi-linear PDEs with FB, Non-variational FB, and Non-local problems. As the techniques for treating the classical obstacle problem substantially differs from the others we have chosen to only discuss these problems from a heuristic point of view, and focus on approaches and the difficulties of these problems.

5.1 Introduction

5.1.1 Background

Our objective is to present several problems within applied sciences, having *Free Boundaries* (FB) as a common denominator. The term Free Boundary Problems (FBP) refers to a class of problems in which one or several variables must be determined in different and a priori unknown domains of the space, or space-time. Hence, finding the domain is part of the problem. The process is usually controlled by several underlying mathematical conditions that are derived from certain physical laws or other constraints governing a phase transition.

The problems we have in mind arise in various mathematical models encompassing applications that ranges from physical to economics, financial and biological phenomenon, where there is an extra effect of the medium. This effect is in general a qualitative change of the medium and hence an appearance of a phase transition: ice to water, liquid to crystal, buying to selling (assets), active to inactive (biology),

H. Shahgholian (✉)
Department of Mathematics, Royal Institute of Technology, Stockholm, Sweden
e-mail: henriksh@kth.se

© Springer Nature Switzerland AG 2018

C. Bianchini et al. (eds.), *Geometry of PDEs and Related Problems*,
Lecture Notes in Mathematics 2220, https://doi.org/10.1007/978-3-319-95186-7_5

blue to red (coloring games), disorganized to organized (self-organizing criticality),[1] etc. The boundary between these two states are called free boundaries, and are a priori unknown. This naturally introduces a substantial difficulty for mathematical analysis of such problems.

Example 5.1 Consider the following partial differential equation

$$\Delta u = 1 \qquad \text{in } B_r(0) \qquad u = 0 \quad \text{on } \partial B_r(0), \tag{5.1}$$

where $B_r(0)$ is the ball with center at the origin, and radius r, and Δ is the Laplace operator. The explicit solution is given by $u = (|x|^2 - r^2)/2n$.

Now suppose we impose a further condition, say any of these:

- $\int u = \lambda_1$, with λ_1 given constant,
- $|\nabla u| = \lambda_2$ on ∂B_1, where $\lambda_2 > 0$.

Then we see that the problem cannot be solved, unless λ_1, or λ_2 are taken appropriately.

An alternative solution to the problem is to change r to accommodate any of the above extra conditions. Changing r, amounts to changing the domain where the equation applies. In other words, if we impose any further condition than the Dirichlet data, then we have to let r (i.e., the domain) in Eq. (5.1) be free.

This is a simple example of a free boundary problem, where extra data is imposed, making the problem overdetermined.

5.1.2 Overview of the Content

In this semi-expository note I shall discuss both at heuristic and technical levels (some of the) tools and theories that have been developed to study free boundary problems. Below is a list of specific problems that I shall discuss:

- *A Catalog of semi-classical FBPs*: Modeling, Applications, Academic questions.
- *Mathematical Theory for obstacle type problems*: Existence, optimal regularity, non-degeneracy, Hausdorff dimension of FB, Monotonicity formulas, Global solutions, Lipschitz and C^1- regularity of FB.
- *Other type of FBPs*: Bernoulli type FBP, Semi- and Quasi-linear PDEs with FB, Non-variational FB, Non-local problems.
- *Recent trends, and Developments*: Switching problems, Constraint energy minimizing maps, and nonlocal and thin obstacles.
- *Open problems and questions.*

---

[1]An interesting aspect of such a criticality is the so-called sandpile dynamic (or Internal DLA), that has been subject to intense studies in the last decade.

5.2 A Catalog of Semi-classical FBPs

5.2.1 A Melting Ice Block

Consider a big piece of ice block, in the middle of Ocean. The heat transfers from water to ice, and melts it gradually. There are two governing PDEs, one in the water region and one in the ice region. The equations depend on the material of water and ice. On the boundary between water and ice region there is the so-called *mushy region*, which is a mixture of both materials. At this region a phase-transition takes place. In this model, the ice melts and turns into water and has a qualitative change in material; hence the name phase transition. In the mushy region (if it is taken as sharp boundary) there is the so-called latent heat, which is described by an equation of lower order, than the bulk equation. In our case it is simply the jump of the normal derivatives of the solution function (that represents heat) u (seen from water and ice region)

$$|\nabla u^+|^2 - |\nabla u^-|^2 = \lambda^2 > 0 \qquad \text{on } \partial\{u > 0\}.$$

In the rest of the domain we have $\Delta u^\pm = 0$ (or some PDE, depending on the ingredients).

5.2.2 Hele-Shaw Flow

Several flow problems give rise to free boundaries: Hele-Shaw, Muskat, Stokes, flows are a few to mention. For definiteness we consider Hele-Shaw flow, which is a flow between two parallel flat plates separated by an infinitesimally small gap. This a mathematical model in micro-lubrication, and thin-film production in industry, see [17], and [16] and the references therein.

Here, I shall touch upon the classical Hele-Shaw problem, and describe the model. Hele-Shaw flow amounts to moving of an initial interface with respect to pressure, in the system. Specifically we consider pressure from the Green's function with a source $\mu \geq 0$ (usually Dirac source)

$$\Delta G^t = -\mu \quad \text{in } D^t, \qquad G^t = 0 \quad \text{on } \partial D^t,$$

where $D^0 \subset \mathbb{R}^n$ ($n \geq 2$) is given and evolves with time, and $\text{supp}(\mu) \subset D^0$. The flow is governed by $-\nabla G^t$, which represent pressure in the outward normal direction, i.e. $V = -\partial_\nu G$ is the speed in the normal direction. Integrating the Green's function in time

$$u(x) = u(x, t) = \int_0^t G^\tau d\tau$$

one obtains a new function u that solves

$$\Delta u = \chi_{D^t} - \chi_{D^0} - t\mu .$$

More importantly u admits a variational formulation, and we can talk about weak solutions. To see how we obtain the PDE above, we work with

$$v(x,t) := \mathscr{F} \star (\chi_{D^t} - t\mu)(x),$$

where $\mathscr{F}$ is the (normalized) fundamental solution, $\Delta\mathscr{F} = \delta_0$. Here '$\star$' denotes convolution. Differentiating v in t we arrive at

$$\frac{d}{dt}v(x,t) = \mathscr{F} \star (V d\sigma - \mu) = \mathscr{F} \star (-\partial_v G^t d\sigma - \mu) = \mathscr{F} \star (\Delta G^t)(x) = G^t(x).$$

In particular $\frac{d}{dt}v(x,t) \geq 0$ with equality if $x \notin D^t$. Here $d\sigma$ is the surface measure, and smoothness is assumed.

Since also $\frac{d}{dt}u(x,t) = G^t(x)$, we have $\frac{d}{dt}(u(x,t) - v(x,t)) = 0$. Integrating, we have (for some $C(x)$ constant in t) $u(x,t) = v(x,t) - C(x)$. Since

$$v(x,0) = \mathscr{F} \star \chi_{D^0},$$

and $u(x,0) = 0$ (by definition) we have $C(x) = \mathscr{F} \star \chi_{D^0}$ and as claimed,

$$u(x,t) = \mathscr{F} \star (\chi_{D^t} - t\mu)(x) - \mathscr{F} \star \chi_{D^0}.$$

Summing up we have that the Hele-Shaw flow problem defines a FBP, where the FB separates the flow and non-flow region.

5.2.3   An Optimal Stopping Problem

The Hele-Shaw problem can be modeled as an optimal stopping problem as follows. For μ as in the previous section, define

$$U_\theta(x) = U_{\theta,t\mu,D^0}(x) := \mathbb{E}^x \left(-\theta + \int_0^\theta \left(t\mu(X_s) + \chi_{D^0}(X_s) \right) ds \right),$$

where X_s is Brownian motion, and θ is any finite valued stopping time.[2] Then the function

$$u(x) := \sup_\theta U_\theta(x)$$

[2]Here X_s is Brownian motion, starting at $X_0 = x$, and the stopping time θ is a non-negative measurable function with respect to sigma algebra generated by X_s, for all s.

Fig. 5.1 Equilibrium state of
the membrane over the
obstacle!
Extracted from book by
Petrosyan-Sh-Uraltseva

solves the geometric flow problem described above. Observe that if $\theta \equiv 0$, then
$U_\theta(x) = 0$, and hence $u(x) \geq 0$.

A closely related and basic problem, is the complementary (variational) problem

$$\min(-\Delta v, v - \psi) = 0, \quad \text{in } D \qquad v = g \text{ on } \partial D$$

where ψ is a given obstacle, with $g \geq \psi$ on ∂D (Fig. 5.1). This in turn relates to the
optimal stopping problem

$$v(x) := \sup_\theta \mathbb{E}^x \left(\psi(X_{\theta \wedge \tau}) \right),$$

where θ is any bounded stoping time, and $\tau = \tau(x)$ is exit time[3] for D.

5.2.4   Modeling Financial Derivatives

Financial instruments (such as options) may give some advantages (such as
possibility of early exercise/callback) to buyers/writers of the contracts. In this case
an unknown exercise region[4] appears, where the boundary of the region is a FB.
Examples of such models appear in

(1) Optimal stopping problems, American type options: Put, call (with dividend),
(2) Problems with Jump diffusion, Min option, Butterfly options,
(3) Transactions costs, Callable contracts, Convertible bonds, Capped options,
(4) Optimal switching problems (switching from one state to another).

---

[3] The first time when the underlying process X_t exits the domain D.

[4] For contracts with possibility of early exercise, the set of values of the underlying
stock/commodity that theoretically is a region when one should exercise the right, is called exercise
region.

These financial instruments correspond to the mathematical models that are related to the following problems:

(1) Variational Inequalities, elliptic/parabolic problem,
(2) Thin/Rooftop/Discontinuous obstacles,
(3) Double obstacle problem, Presence of fixed boundary,
(4) System of inter-obstacles.

Let us derive the mathematical model of pricing financial derivatives, and relate it to the obstacle problem. To start with, we know that arbitrage theory tells us the no-free lunch principle:

$$\boxed{\text{Return from Bank} \geq \text{Return from the Hedged portfolio}}$$

Let V be the value of an option for an underlying asset with value S. Suppose Π is a portfolio with one option long and m assets[5] short. Then we have

$$\Pi = V - mS, \qquad d\Pi = dV - mdS.$$

This, along with $dS = \sigma S dW + r S dt$ and Ito's Taylor-expansion formula[6] and the choice of $m = dV/dS$ leads to

$$V_t + \frac{1}{2}\sigma^2 S^2 V_{SS} + rSV_S - rV \leq 0, \tag{5.2}$$

along with a further important ingredient $V \geq \psi$. I.e. a guarantee that the price will never fall below the value ψ, a given function. A last ingredient (to buyer's benefit) is that V should be the smallest possible value! Otherwise another writer will offer a smaller value and still make profit. In other words, the smallest "super-solution" to the PDE above. Hence we conclude that financial instruments with extra benefits are modeled by the variational inequality:

Find smallest $V \geq \psi$ satisfying inequality (5.2).

5.2.5 Smash Sums and Internal DLA

(See [10]) The smash sum C of two sets $A, B \subset \mathbb{Z}^2$ is a certain random set defined as follows. Let $A \cap B = \{x_1, \cdots, x_k\}$. Each point in $A \cap B$, is sent out through random walks. Once the point hits *first time* any point y_j outside the region $A \cup B$ it stops and adds y_j to the union: $C_j = A \cup B \cup \{y_j\}$.

---

[5] Assets are assumed divisible, so that $m \in \mathbb{R}$.
[6] (using $dW^2 = dt$).

Fig. 5.2 Smash sum of two rectangles (Courtesy of H. Aleksanyan)

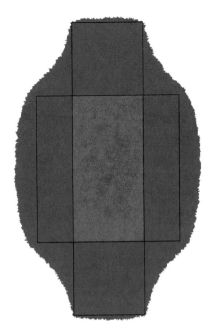

The process continuous until all points x_j are sent out. The resulting region $C :=C_k$ is the Smash sum, (Fig. 5.2) and is independent of the order of x_j. As the lattice spacing tends to zero, the smash sum has a deterministic scaling limit, call it u. This solves an obstacle type problem

$$\Delta u = \chi_{\{u>0\}} - 2\chi_{A \cap B}, \qquad u \geq 0.$$

This is also related to the so-called Quadrature Domains; see [15].

5.3   Mathematical Theory of Obstacle Problem

5.3.1   Existence, and Uniqueness

Existence theory for (scalar) variational problems are well developed. One can use many different approaches, depending on the problem. For the obstacle problem that we are involved with in this note one has the following approaches: penalization, Perron's method, minimization, variational inequality (projection), iterative scheme; the latter is good for numerics, and system case.

Let us now consider the penalization approach, [14]. For this purpose we consider a solution $v_{\epsilon,k}$ to the PDE

$$\Delta v_{\epsilon,k} = \beta_{\epsilon,k}(v_{\epsilon,k} - \psi) \qquad \text{in } B_1,$$

where B_1 is the unit ball, and $\beta_{\epsilon,k}(t)$ can be taken as

$$\beta_{\epsilon,k}(t) = \epsilon \min(1, t^+) - \min(\frac{t^-}{\epsilon}, k),$$

and the boundary values $v_{\epsilon,k} = g(x) \geq \psi$ on ∂B_1. One may use fixed point theory to solve this problem, for each ϵ, k; observe that all ingredients are smooth here.

Now a clever use of minimum principle[7] can help to show

$$-C \leq \beta_{\epsilon,k}(v_{\epsilon,k} - \psi) \leq \epsilon,$$

i.e.

$$|\Delta v_{\epsilon,k}| \leq C, \qquad \text{independent of } \epsilon, k,$$

which, by elliptic estimates, gives uniform $W_{loc}^{2,p} \cap C_{loc}^{1,\alpha}$-regularity for $v_{\epsilon,k}$. Letting $k \to \infty, \epsilon \to 0$, and using uniform estimates we conclude $v_{\epsilon,k} \to v$, which solves

$$\Delta v \leq 0, \qquad v \geq \psi, \qquad (v - \psi)\Delta v = 0 \quad \text{a.e.}$$

The interpretation of the above problem can be either in the weak sense or viscosity. Also from the last properties of the limit function v for the penalization method, we see that $\Delta v \leq 0$, and $v \geq \psi$. This suggests that Perron's method of smallest super-solution may well be used to find a solution to the problem above. Indeed, the smallest super-solution satisfying the above inequalities give rise to a unique function which we call a solution to the obstacle problem.

Next, we discuss the minimization problem

$$\min_{v \in \mathcal{K}} \int_{B_1(0)} |\nabla v|^2 dx, \tag{5.3}$$

where $\mathcal{K} = W_g^{1,2} \cap \{\psi_1 \leq v \leq \psi_2\}$,   (double obstacle)

and g is boundary data, ψ_1 is the lower obstacle, and ψ_2 the upper obstacle. For $\psi_2 = \infty$, this is just the one-sided obstacle problem. More generally one can allow terms of type $F(x, v)$ in the above integral, with F convex in v, and $\mathcal{K}$ a suitable convex space.

---

[7]Since all ingredients are smooth the solution is in $W^{2,p}$, and hence in $C^{1,\alpha}$. From here it follows that $h(x) = \beta_{\epsilon,k}(v_{\epsilon,k}(x) - \psi(x))$ is Lipschitz (since $\beta_{\epsilon,k}$ is Lipschitz). By elliptic PDE $v_{\epsilon,k}$ is C^2. Now the function $h(x)$ takes a minimum at some point x^0, which by monotonicity of β implies that $v_{\epsilon,k} - \psi$ takes minimum at x^0, and hence $\Delta(v_{\epsilon,k} - \psi) \geq 0$, implying $\beta_{\epsilon,k}(v_{\epsilon,k} - \psi)(x^0) = \Delta v_{\epsilon,k}(x^0) \geq \Delta \psi(x^0) \geq -C$. In other words, $\Delta v_{\epsilon,k} = \beta_{\epsilon,k}(v_{\epsilon,k} - \psi)$ is universally bounded, and hence $C^{1,\alpha}$.

Due to lower semi-continuity of the functional we always have minimizers. It is not hard to see that any minimizer v will be unique. Suppose v_1, v_2 are minimizers, with $\nabla v_1 \neq \nabla v_2$, then

$$\int_{B_1} \left| \nabla \left(\frac{v_1 + v_2}{2} \right) \right|^2 < \int_{B_1} \frac{1}{2} |\nabla v_1|^2 + \int_{B_1} \frac{1}{2} |\nabla v_2|^2 \leq \int_{B_1} \left| \nabla \left(\frac{v_1 + v_2}{2} \right) \right|^2$$

which is a contradiction, unless $\nabla v_1 = \nabla v_2$, a.e., and hence $v_1 = v_2 + c$ in L^2. Since they have the same boundary values, we must have $c = 0$ and hence $v_1 = v_2$.

It is not hard to see that any minimizer v will be harmonic in the set $\psi_1 < v < \psi_2$; since then one can make variations both upwards and downwards through $v \pm \epsilon \varphi$, with ϵ small. Therefore our minimizer will satisfy $\psi_1 \leq v \leq \psi_2$, and (assuming ψ_i are smooth enough, say C^2)

$$\Delta v = (\Delta \psi_1) \chi_{\{int(v=\psi_1)\}} + (\Delta \psi_2) \chi_{\{int(v=\psi_2)\}} + \mu \tag{5.4}$$

where (as we'll show below) μ is a (signed) measure supported on the free boundary, i.e., where the solution detaches from the obstacle

$$support(\mu) \subset \partial \{\psi_1 < v < \psi_2\}.$$

We will show that $\mu \equiv 0$, which means that the solution detaches smoothly from the obstacle.

For simplicity we assume $\psi_2 = +\infty$, so we only have one obstacle, i.e the one from below. A similar reasoning will work for the upper obstacle. For points in between both obstacles one naturally has the same smoothness as the obstacles. For some recent work on double obstacle problem we refer to [20]. For $u := v - \psi_1$ we have (in the weak sense)[8]

$$\Delta u = f \chi_{\{u>0\}} + \mu, \qquad u \geq 0, \tag{5.5}$$

where $f(x) = -\Delta \psi_1(x)$ and $support(\mu) \subset \partial \{u > 0\}$. Let $\eta \in C_0^\infty(B_1)$, $\eta \geq 0$. For $\epsilon > 0$ define

$$\eta_\epsilon = \eta \chi_\epsilon,$$

---

[8] Observe also that for the complementarity problem

$$\Delta v \leq 0, \qquad v \geq \psi, \qquad (v - \psi) \Delta v = 0 \quad \text{a.e.}$$

we also have $\Delta v = \Delta \psi \chi_{\{int(v=\psi)\}} + \mu$.

where

$$\chi_\epsilon = \begin{cases} 1 & \text{if } u(x) \geq 2\epsilon \\ \frac{u(x)}{\epsilon} - 1 & \text{if } \epsilon < u(x) < 2\epsilon \\ 0 & \text{if } u(x) \leq \epsilon. \end{cases}$$

This is meaningful, since u is continuous when $u > 0$. Set $\Omega_+ = \{u > 0\}$[9] then we have

$$-\langle \eta_\epsilon, f \chi_{\Omega_+} \rangle = \int_{\Omega_+} \nabla u \cdot \nabla \eta_\epsilon \, dx = \int_{B_1} (\nabla u \cdot \nabla \eta) \chi_\epsilon \, dx$$

$$+ \frac{1}{\epsilon} \int_{\epsilon < u < 2\epsilon} |\nabla u|^2 \eta \, dx \geq \int_{B_1} (\nabla u \cdot \nabla \eta) \chi_\epsilon \, dx.$$

Since $0 \leq \eta_\epsilon \leq \eta$ and $\int_{B_1} |\nabla u||\nabla \eta| \, dx < \infty$, we can let $\epsilon \to 0$, to obtain (through dominated convergence theorem)

$$-\langle \eta, f \chi_{\Omega_+} \rangle \geq \int_{\Omega_+} \nabla u \cdot \nabla \eta \, dx = \int_{B_1} \nabla u \cdot \nabla \eta \, dx.$$

Here we have used that $\nabla u = 0$ a.e. in $B_1 \setminus \Omega_+$. The last inequality is equivalent to $\mu \geq 0$, in the Eq. (5.5).

Make a variation upwards,[10] with $0 \leq \varphi \in W_0^{1,2}(B_1)$, to obtain

$$\int \nabla u \cdot \nabla \varphi + f \varphi \geq 0 \quad \text{i.e.} \quad \Delta u \leq f \chi_{\overline{\Omega_+}} \qquad \text{in } B_1$$

where we used $u = 0$ in $(\overline{\Omega_+})^c$. From here we have

$$f \chi_{\Omega_+} + \mu = \Delta u \leq f \chi_{\overline{\Omega_+}}$$

which along with $\mu \geq 0$ implies

$$0 \leq \mu \leq f \chi_{\partial \Omega_+} \in L^\infty.$$

Hence

$$\Delta u = h, \qquad u \geq 0, \qquad \text{in } B_1,$$

---

[9]We assume u continuous, to make sure this domain is open. The proof of continuity needs a few lines of argument, that is outside the scope if this note.

[10]It is obvious that if v minimizes (5.3) then u minimizes $\int |\nabla u|^2 + 2fu$.

where

$$f \chi_{\{u>0\}} \le h \le f \chi_{\overline{\{u>0\}}}, \qquad f \in C^\alpha(B_1).$$

Since $|h| \le C$ we have $u \in W^{2,p}_{loc} \cap C^{1,\alpha}_{loc}(B_1)$ (for all $p < \infty$, $\alpha < 1$). In particular $D^2 u = 0$ a.e. in $B_1 \setminus \Omega_+$, and hence $\mu = 0$ a.e. But μ being bounded and zero a.e. it must be zero. It follows now that $h = f \chi_{\{u>0\}}$, and $\mu \equiv 0$.

5.3.2   Optimal Regularity and Non-degeneracy of Solutions

The next question we want to address is the optimal regularity of solutions/minimizers.

- RHS $\in C^\alpha$ $\qquad\qquad \Longrightarrow \qquad$ sol. $C^{2,\alpha}$,
- RHS $\in L^\infty$ $\qquad\qquad \Longrightarrow \qquad$ sol. $C^{1,\alpha}$, any $\alpha < 1$,
- RHS $\in$ *which Space* $\qquad \Longrightarrow \qquad$ sol. $C^{1,1} \setminus C^2$.

For FBPs, it is common that we also obtain the last missing regularity. In this case we will prove that solutions to

$$\Delta u = f \chi_{\Omega_+} \in L^\infty$$

are $C^{1,1}_{loc}$ whenever f is C^α. Henceforth, we assume $f \ge \lambda_0 > 0$, as this is crucial for regularity of FB, but not for the solutions u. This assumption prevents degeneracy of solutions; this is explained later. It means we are only interested in FB points where $\Delta \psi_1 \le -\lambda_0 < 0$. Observe that due to super-harmonicity of minimizers they do not touch parts of the obstacle where $\Delta \psi_1 > 0$. We make a further reduction and assume $f \equiv 1$, i.e. $\psi_1 = l(x) - |x|^2/2n$, with $l(x)$ linear.

We shall now prove the optimal regularity of u in $B_{1/4}(0)$ only, for simplicity. Choose $z \in \partial\{u > 0\} \cap B_{1/2}(0)$, and let $0 < r < 1/3$. Suppose we know that

$$\sup_{B_r(z)} u(x) \le C_0 r^2,$$

for a universal C_0. This will give a quadratic growth/decay for solutions from the FB points. Why is this enough for showing $u \in C^{1,1}$? Indeed, for any point $y \in \{u > 0\} \cap B_{1/4}(0)$, we define, in $B_1(0)$

$$\tilde{u}_r(x) = \frac{u(y + d_y x)}{r^2},$$

where $d_y = dist(y, \{u = 0\})$. Let further $z \in \partial\{u > 0\}$ be (any) closest point to y. Then according to the estimate above we have $0 \le \sup_{x \in B_{2r}(z)} \tilde{u}_r(x) \le 4C_0$. Since also $\Delta \tilde{u}_r = 1$, we can apply elliptic regularity to obtain $|D^2 u(y)| = |D^2 \tilde{u}_r(0)| \leq C_1$. To close the argument, we need to show the above quadratic decay estimate. For this we follow the approach[11] as in [6] for the original obstacle problem; cf. also [4] in the formulation we have here. Decompose the function u as $u = u_1 + u_2$, where

$$\Delta u_1 = 0 \quad \text{in } B_r(z) \qquad u_1 = u \quad \text{on } \partial B_r(z),$$

$$\Delta u_2 = \chi_{\{u>0\}} \quad \text{in } B_r(z) \qquad u_2 = 0 \quad \text{on } \partial B_r(z),$$

and observe that $u(z) = 0$, and $u \geq 0$ in B_1, $u_2 \leq 0$, and $u_1 \geq 0$. In particular Harnack's Inequality applies to u_1 and we conclude $u_1(x) \leq c u_1(z)$ for $x \in B_{r/2}(z)$, with $c > 0$. Putting things together we have

$$u(x) = u_1(x) + u_2(x) \leq u_1(x) \leq c u_1(z) = c(u(z) - u_2(z)) = -c u_2(z),$$

where we have used that $u_2(x) \leq 0$, and $u(z) = 0$. So it remains to show $-u_2(z) \leq Cr^2$. To this end we use a barrier $h(x) = (|x - z|^2 - r^2)/2n$, and conclude by the comparison principle that $u_2(x) \geq h(x)$ in $B_r(z)$ and hence $u_2(z) \geq -r^2/2n$, giving us $u(x) \leq -c u_2(z) \leq c_1 r^2$, which is the desired result. Hence we have $u \in C^{1,1}$. The above argument actually shows that

$$u(x) \leq c_1 (dist(x, \partial \Omega_+))^2. \tag{5.6}$$

What is next step? Can a solution converge faster to zero than quadratic? And if so, how does it affect our analysis? Indeed, if $\Delta u = f$, and $f(z) = 0$, then we have a good chance to prove a faster than quadratic decay for u at z.[12] This is just elliptic regularity. Analysis of such points fall outside the regularity theory developed by L. Caffarelli for obstacle problem. So we shall prove that for the case $f > \lambda_0 > 0$, or as we have assumed $f \equiv 1$, the solution has optimal decay r^2, and not faster.

We shall prove that for any FB point $z \in \partial\{u > 0\}$ we have $\sup_{B_r(z)} u \geq r^2/2n$. This follows from comparison with $h(x) = |x - z|^2/2n$. Indeed if $h \geq u$ on

---

[11]There are several other ways of doing this, for this particular problem. Two such methods are the use of mean value theorem along with Schwarz potential (see [22] Exercise 2.1), and the second one is the blow-up technique. A different way of seeing such regularity is to consider the smallest super-solution for $\Delta v \leq 1$, $v \geq 0$, and $v = g$ on the boundary ∂B_1. Then $v(x) = [u(x + he) + u(x - he)]/2 + C|h|^2$ is also a super-solution and for large enough C it satisfies $v \geq g$ on the boundary, provided g is C^2. In particular $v \geq u(x)$, since u is the smallest super-solutions. This in turn implies u is $C^{1,1}$ from below. But since Δu is bounded we conclude u is $C^{1,1}$. This argument has been used for the thin obstacle problem to prove almost convexity of solutions by L. Caffarelli and collaborators.

[12]This statement is true, at least if $f(x)$ is Dini at z. One should however be careful as there are examples of type $x_1 x_2 (-\log |x|)^a$, with $0 < a < 1$ that does not even give us quadratic growth at the origin.

$\partial B_r(z)$, then by comparison principle[13] for FBP we have $h \geq u$. In particular if $u(z) > 0$, then $0 = h(z) \geq u(z) > 0$, which is a contradiction. Hence $\sup_{B_r(z)} u \geq \sup_{B_r(z)} h(x) = r^2/2n$, and this holds in the limit as $z \to$ FB. This nice argument I picked up in G.S. Weiss' paper [23].

So far we have shown that for $z \in \partial \Omega_+ = \partial\{u > 0\}$

$$c_0 r^2 \leq \sup_{\partial B_r(z) \cap \Omega_+} u \leq c_1 r^2 \qquad (c_0, c_1 \text{ universal})$$

and that also $u \in C_{loc}^{1,1}(B_1)$. Why did we insist on finding these extremes estimates on both sides, and how will we be using them later on? The reason is to find the invariant scaling of our equation. Indeed if we look at the scaling

$$u_r(x) := \frac{u(rx + z)}{r^a} \qquad \text{(for some } a > 0\text{)}$$

then we see that $\Delta u_r = r^{2-a} \chi_{\{u(rx)>0\}} = r^{2-a} \chi_{\{u_r(x)>0\}}$. Hence to keep the equation invariant under scaling, we need $a = 2$, so that we can retain the original equation, during any scaling.

Now after ensuring that the equation is invariant under this scaling, we also need to make sure, that the scaled function $u_r = u(rx + z)/r^2$ (at FB point z) is:

(1) Uniformly bounded on compact sets of $\mathbb{R}^n$.
(2) Does not flat out to zero, in the limit.

These are guaranteed by optimal growth and non-degeneracy as we showed above. Let us now discuss properties of FB.

- Can it have positive measure?
- Is $(n-1)$-Hausdorff measure of FB finite?

We recall the fact that the Lebesgue upper density of a set E must be 1 for a.a. points of E:

$$\limsup_{r \to 0} \frac{|E \cap B_r(x^0)|}{|B_r|} = 1, \qquad \text{for a.a. } x^0 \in E.$$

We shall use this along with the optimal quadratic growth and non-degeneracy to conclude the FB has zero Lebesgue measure.

Let $z \in \partial \Omega_+$, and take $r > 0$ (not too big). Then we know that on $\partial B_{r/2}(z)$ there is a point y_r such that $c_0(r/2)^2 \leq u(y_r) \leq c_1(dist(y, \partial \Omega_+))^2$, where the second inequality uses optimal quadratic decay. This means that for any r, and $B_r(z)$ (as

---

[13]Observe that this is not the standard comparison principle for PDEs, but for FBPs, that shows large boundary data gives larger solutions. Here one has to see h as a solution to the obstacle problem in $B_r(z)$ with boundary values $r^2/2n$.

above) we have a point $y \in B_{r/2}(z)$ such that $dist(y_r, \partial\Omega_+) \geq c_2 r$. In other words $B_{c_2r}(y_r) \subset \Omega_+ \cap B_r(z)$. This means that for any $z \in \partial\Omega_+$

$$\frac{|\partial\Omega_+ \cap B_r(z)|}{|B_r|} = \frac{|\partial\Omega_+ \cap (B_r(z) \setminus B_{c_2r}(y_r))|}{|B_r|} \leq 1 - c_3 < 1.$$

But according to Lebesgue density theorem this set is negligible.

We shall now prove that $\partial\Omega_+(u)$ is a set of finite $(n-1)$-dimensional Hausdorff measure[14] locally in $B_{1/2}$. I.e., we want to cover the FB with $(1/\epsilon)^{n-1}$-balls of radius ϵ, and take infimum as ϵ tends to zero. Let

$$v_i = \partial_{x_i} u, \quad i = 1, \dots, n, \qquad E_\varepsilon = \{|\nabla u| < \varepsilon\} \cap \Omega_+.$$

Observe that $1 \leq |\Delta u|^2 \leq c_n \sum_{i=1}^{n} |\nabla v_i|^2$ in Ω_+. Thus, for an arbitrary compact set $K \Subset B_1$ we have

$$|K \cap E_\epsilon| \leq c_n \int_{K \cap E_\epsilon} \sum_i |\nabla v_i|^2 dx \leq c_n \sum_i \int_{K \cap \{|v_i| < \epsilon\} \cap \Omega_+} |\nabla v_i|^2 dx.$$

To estimate the right hand side, we notice that

$$\Delta v_i^\pm = \Delta(\partial_{x_i} u)^\pm \geq 0, \qquad \text{i.e.,} \qquad \int_{B_1} \nabla v_i^\pm \nabla\eta \, dx \leq 0, \quad i = 1, \dots, n$$

for any non-negative $\eta \in C_0^\infty(B_1)$, and by continuity, for any nonnegative $\eta \in W_0^{1,2}(B_1)$. If we now choose $\eta = \psi_\epsilon(v_i^\pm)\phi$, with

$$\psi_\varepsilon(t) = \begin{cases} 0, & t \leq 0 \\ \epsilon^{-1}t, & 0 \leq t \leq \epsilon \\ 1, & t \geq \epsilon \end{cases}$$

and a nonnegative cutoff function $\phi \in C_0^\infty(B_1)$, $\phi = 1$ on K, we will obtain

$$\int_{B_1} \nabla v_i^\pm \nabla(\psi_\epsilon(v_i^\pm)\phi) dx = \int_{\{0 < v_i^\pm < \epsilon\}} \epsilon^{-1} |\nabla v_i^\pm|^2 \phi \, dx$$

$$+ \int_{B_1} \nabla v_i^\pm \psi_\epsilon(v_i^\pm) \nabla\phi \, dx \leq 0$$

---

[14]The argument here is due to Caffarelli [5], and as to the writing of this paper the author knows of no other approach for Hausdorff measure estimate for obstacle type problems. Indeed, the problem is open for general operators.

which implies that

$$\epsilon^{-1} \int_{K\cap\{|v_i|<\epsilon\}\cap\Omega_+(u)} |\nabla v_i|^2 dx \le \epsilon^{-1} \int_{\{0<|v_i|<\epsilon\}} |\nabla v_i|^2 \phi\, dx$$

$$\le \int_{B_1} |\nabla v_i||\nabla\phi|\, dx \le c_n M \int_{B_1} |\nabla\phi|\, dx,$$

where $M = \|D^2 u\|_{L^\infty(B_1)}$. Thus, summing over $i = 1, \ldots, n$, we will arrive at an estimate $c_0 |K \cap E_\epsilon| \le C\epsilon M$, where $C = C(n, K, B_1)$.

Consider now a covering of $\partial\Omega_+ \cap K$ by a finite family $\{B^i\}_{i\in I}$ of balls of radius ϵ centered on $\partial\Omega_+ \cap K$, such that no more than $N = N_n$ balls from this family overlap (Besicovitch covering). For $\epsilon > 0$ sufficiently small, we may assume that $B^i \subset K'$ for a slightly larger compact K' so that $K \Subset \text{Interior}(K') \Subset B_1$. Now notice that $|\nabla u| < M\epsilon$ in each B^i, implying that

$$B^i \cap \Omega_+ \subset E_{M\epsilon} = \{x : |\nabla u| < M\epsilon\}.$$

Then, using density argument and estimate above, we obtain

$$\sum_{i\in I} |B^i| \le C_1 \sum_{i\in I} |B^i \cap \Omega_+| \le C_1 \sum_{i\in I} |B^i \cap E_{M\epsilon}| \le N C_1 |K' \cap E_{M\epsilon}|$$

But we have already shown that $c_0 |K \cap E_\epsilon| \le C\epsilon M$, which combined with the above estimate (with ϵ replaced by $M\epsilon$) implies $\sum_{i\in I} |B^i| \le \frac{C_1 C N M^2 \epsilon}{c_0}$. This gives the estimate

$$\sum_{i\in I} \text{diam}(B^i)^{n-1} = \sum_{i\in I} \epsilon^{n-1} \le C(n, M, K, B_1),$$

which by letting $\epsilon \to 0$, and taking infimum, implies

$$H^{n-1}(FB) \cap K) \le C(n, M, K, B_1).$$

5.3.3   Regularity of FB: Local and Global Analysis

Since FB in our problems are boundary of a set, and as such they are regular if locally they can be as graph of regular functions. Now how to prove something is locally a (smooth) graph? Observe that the FB is $\partial\{u > 0\}$. So we need somehow to involve the function u in our analysis. As an example consider (for $m > 0$)

$$u(x) := \left[(x_2 - \cos m x_1 + 1)_+\right]^2.$$

Then $\Delta u = m^2 f(x) \chi_{\{u>0\}}$, for some smooth f, with $f(0) > 0$. Moreover for $e = (a_1, a_2) \approx e_2$,

$$D_e u \approx c_1 a_2(x_2)_+ \approx c_2 a_2 \sqrt{u} \geq c_3 a_2 u \geq 0,$$

if we are close to FB. Hence we have $\partial\{u > 0\}$ is a graph in e-direction. Since e can vary (at least slightly) we may deduce Lipschitz regularity of FB. This suggests that for $e \approx e_2$, if we are close to FB, we need to prove $D_e u - cu \geq 0$.

Now if we take m larger, then the approximate neighbourhood for the inequality above will be smaller. This calls for some care, when proving regularity of FB. So far we have learned that solutions of one of our problems, which in its simplest form solves $\Delta u = \chi_{\{u>0\}}$, $u \geq 0$, in B_1, have some good properties such as for $\Gamma = \partial\{u > 0\} = \partial\Omega_+$

$$c_0 r^2 \leq \sup_{B_r(z)} u(x) \leq c_1 r^2, \qquad z \in \Gamma \cap B_{1/2}.$$

These bounds give us that the scaled function $u_r(x) = \frac{u(rx+z)}{r^2}$ satisfies $\Delta u_r = \chi_{\{u_r>0\}}$, $u_r \geq 0$, in $B_{1/r}$, and $c_0 \leq \sup_{B_1(0)} u_r(x) \leq c_1, 0 \in \Gamma_r$. Here $\Gamma_r = \frac{1}{r}(\Gamma - z)$ is scaled version $\Gamma - z$.

The last two conditions guarantee that u_r (for a subsequence) will converge to a solution in $\mathbb{R}^n$, for the same problem $\Delta u_0 = \chi_{\{u_0>0\}}$, $u_0 \geq 0$, in $\mathbb{R}^n$, and moreover $\sup_{B_R} u_0(x) \leq C R^2 \ \forall \ R > 0$. To see the latter we set $x = Ry$, with $|y| = 1$. Then for some subsequence r_j

$$u_0(x) \approx \frac{u(r_j x + z)}{r_j^2} = \frac{u(r_j R y + z)}{(R r_j)^2} R^2 \leq c_1 R^2.$$

What does this tell us? Suppose the FB, Γ at point z has some irregularities, say a cusp, or a cone shape singularity. If so, then this property must be reflected in the scaled version Γ_r, and hence for the blow-up Γ_0 (Fig. 5.3). On the other hand if Γ at point z is regular, say C^1, then Γ_0 must be a hyperplane, with $u_0 > 0$ on one side and $u_0 = 0$ on the other side.

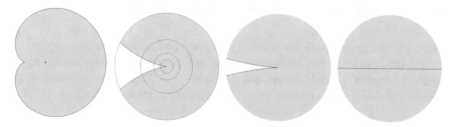

Fig. 5.3 The Cardioid from left to right scaled at its cusp point

The above pictures suggests that the local analysis of FB has to take into consideration the global pictures, after scaling, and blow-up. Therefore we need to see what kind of global solutions we shall have after blow-up of a solution. We shall first make a reasonable observation:

> Blow-ups are homogeneous.

To prove homogeneity one uses the Weiss-monotonicity function[15]

$$W(r, u, x^0) := \int_{B_1(0)} (|\nabla u_r|^2 + 2u_r)\, dx - 2 \int_{\partial B_1(0)} u_r^2 \, dH^{n-1},$$

where $u_r(x) := \frac{u(rx + x^0)}{r^2}$, and $x^0 \in \partial\{u > 0\}$. A simple computation shows

$$W'(r) = \frac{2}{r} \int_{\partial B_1(0)} (u_r')^2 \, dH^{n-1},$$

which is strictly positive unless $u_r' = 0$. Observe that $u_r' = 0$ for $r_1 < r < r_2$ implies that for $x \in B_{r_2} \setminus B_{r_1}$ and $y = rx + x^0$

$$rx \cdot \nabla u(rx + x^0) - 2u(rx + x^0) = (y - x^0) \cdot \nabla u(y) - 2u(y) = 0.$$

This means u is homogeneous of degree 2 at the point x^0. Now we see that for $s > 0$

$$W(0^+, u, x^0) = \lim_{r \to 0} W(sr, u, x^0) = \lim_{r \to 0} W(s, u_r, 0) = W(s, u_0, 0)$$

and hence $W(s, u_0, 0)$ is constant in s, i.e. $W' = 0$, which implies that the blow-up u_0 is homogeneous of degree 2. The conclusion from above is that a blow-up is homogeneous function of degree 2. So it remains now to classify homogeneous global solutions of degree 2. Our next task will be to show:

> Any 2-homogeneous solution is either a polynomial or 1-dim.

The only one dimensional solutions (after suitable rotations) are either 2-degree polynomials or $u_0 = \frac{1}{2}(x_1^+)^2$. To prove the above classification, we can look at two possible situations

(1) The set $\{u_0 = 0\}$ has empty interior.
(2) The set $\{u_0 = 0\}$ has non-void interior.

---

[15]This monotonicity formula has been known in elliptic PDEs, in connection to Harmonic maps. It has also been successfully applied to study singular sets in elliptic PDEs, as well.

The first case, along with the fact that the FB has zero Lebesgue measure, implies that $\Delta u_0 = 1$ a.e. in $\mathbb{R}^n$, and moreover it has quadratic growth. By Liouville's theorem we obtain u_0 is a polynomial of degree two.

The second case is more complicated, and one may try first to prove convexity of u_0, i.e., to prove $D_{ee}u_0 \geq 0$ for any direction e. Hence we claim

$$\boxed{\text{For any global solution } u_0 \text{ we have } D_{ee}u_0 \geq 0}$$

If the claim fails, then since u_0 is $C^{1,1}$, we must have $D_{ee}u_0(x) \geq -C > -\infty$ and hence there is a minimizing sequence x^j such that

$$D_{ee}u_0(x^j) \quad \rightarrow \quad -m = \inf_{\mathbb{R}^n} D_{ee}u_0(x) > -\infty.$$

Now let $d_j = \text{dist}(x^j, \{u_0 = 0\})$, and set $v_j(x) := \frac{u_0(d_j x + x^j)}{d_j^2}$. Then $\Delta v_j = 1$ in B_1, and v_j are uniformly bounded (due to quadratic growth of u_0). Hence a subsequence of v_j converges to a new global solution v_0 satisfying $\Delta v_0 = 1$ in B_1, and moreover

$$D_{ee}v_0(0) = -m, \quad \text{and} \quad D_{ee}v_0(x) \geq -m.$$

By minimum principle $D_{ee}v_0(x) = -m$ in the connected component of $\{v_0 > 0\}$ that contains the origin, which we call Ω_0 and may consider the problem only in this domain, and redefine v_0 as zero outside this domain. Let us rotate and set $e = e_1$, and set $D_{11}v_0(x) = -m$.

W.l.o.g. we assume $e = e_1$. Set $D_{11}v_0(x) = -m$, and integrate

$$v_0(x) = -mx_1^2/2 + a(x')x_1 + b(x'), \quad \text{in } \Omega_0.$$

Since for each $z \in \Omega_0$ we have that $v_0(z \pm t_\pm e_1) = 0$ for some $t_\pm = t_\pm(z)$; otherwise $v_0 < 0$ for large values of $|x_1|$. Fix such an interval, call it $I(z)$. Then $D_1 v_0 = -mx_1 + a$ vanishes on both ends of this interval, which is impossible for a linear function. Hence we conclude that $m = 0$, and $D_{ee}u_0 \geq 0$.

Now we use convexity to show that blow-ups are 1-dimensional. Indeed, the convexity of u_0 implies $\{u_0 = 0\}$ is convex, and moreover $D_e u_0 \geq 0$, along any line ray that emanates from the boundary of Ω_0, in the direction e.

From here, the homogeneity and convexity of u_0, and that $\{u_0 = 0\}$ is a cone with largest angle being less than π. Since $D_1 u_0$ is harmonic, Lipschitz, and non-negative we can use (2-dimensional) barriers to arrive at a contradiction unless $\{u_0 = 0\}$ is a half-space, and u_0 is 1-dimensional, that is half-space solution $(x_1^+)^2/2$.

Let us summarize what we have so far: A local solution can be scaled and blown up, to obtain a homogeneous global solution, which is either a polynomial or a half-space solution $(x_1^+)^2/2$. Examples of the cardioid type as well as simple polynomials $(x_1^2 + x_2^2)/4$ in dimension 3 suggests that we should pay attention to possible singular points, that may appear. Therefore to obtain regularity we shall

Fig. 5.4 Flatness of $\Gamma(u)$

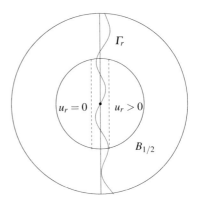

need to have some restriction on the FB-points. E.g. that the free boundary has no cusp at z, i.e., at least one blow-up cannot be a polynomial, and has to be a half-space solution. A way of doing so, is to look at the balanced energy. Let u be a solution of our problem and $x^0 \in \Gamma(u)$. Then we say

$$x^0 \text{ is regular} \quad \text{if} \quad W(0^+, u, x^0) = \frac{\alpha_n}{2},$$

$$x^0 \text{ is singular} \quad \text{if} \quad W(0^+, u, x^0) = \alpha_n,$$

where $\alpha_n = W(r, P, 0) = \cdots = \frac{H^{n-1}(\partial B_1)}{2n(n+2)}$, and $P(x)$ is any polynomial solution to our problem.[16] The conclusion is that if x^0 is a regular point according to the balanced energy, $W(0^+, u, x^0) = \frac{\alpha_n}{2}$, then any blow-up of u at x^0 is a half-space solution. More exactly we obtain that for $\epsilon > 0$ there exists r_ϵ (small enough) such that for any $r < r_\epsilon$ we have

$$\|u_r - u_0\|_{C^{1,\alpha}(B_1)} \le \epsilon \quad \text{where } u_0(x) = \frac{1}{2}(x_n^+)^2.$$

This in turn implies (using non-degeneracy, and that $\limsup\{u_{r_i} = 0\} \subseteq \{u_0 = 0\}$) (Fig. 5.4)

$$u_r > 0 \quad \text{in } \{x_n > \sqrt{2\epsilon}\} \cap B_1, \quad \text{and} \quad u_r = 0 \quad \text{in } \{x_n \le -2\sqrt{n\epsilon}\} \cap B_{1/2}.$$

In particular,

$$\Gamma(u_r) \cap B_{1/2} \subset \{|x_n| \le 2\sqrt{n\epsilon}\}.$$

---

[16] For any order 2 homogeneous solution u, and any degree 2 polynomial P with $\Delta P = 1$ we can use Green's identity to show $\int_{B_1} u = \int_{B_1} P$.

If the free boundary point x^0 is regular (in the sense of energy $W(0^+, u, x^0) = \alpha_n/2$) then a scaled version of our solution u_r satisfies

$$C\partial_e u_r - r u_r \geq C\partial_e u_0 - r u_0 - (C+r)\epsilon \geq -\frac{1}{8n} \quad \text{in } B_1, \tag{5.7}$$

if $e \approx e_1$, or more exactly if

$$e \in \mathscr{C}_\delta := \{x \in \mathbb{R}^n : x_n > \delta|x'|\}, \quad x' = (x_1, \ldots, x_{n-1}),$$

where $\delta > 0$ is small enough. This, we shall prove, implies

$$C\partial_e u_r - r u_r \geq 0 \quad \text{in } B_{1/2}. \tag{5.8}$$

Assume w.l.o.g. $r = 1$ and suppose (despite (5.7)) the inequality (5.8) fails. Let $y \in B_{1/2} \cap \Omega$ be such that $C\partial_e u(y) - u(y) < 0$. Consider then the auxiliary function

$$w(x) = C\partial_e u(x) - u(x) + \frac{1}{2n}|x - y|^2.$$

It is easy to see that w is harmonic in $\Omega \cap B_{1/2}(y)$, $w(y) < 0$, and that $w \geq 0$ on $\partial\Omega$. Hence by the minimum principle, w has a negative infimum on $\partial B_{1/2}(y) \cap \Omega$, i.e.

$$\inf_{\partial B_{1/2}(y)\cap\Omega} w < 0, \quad \text{i.e.,} \quad \inf_{\partial B_{1/2}(y)\cap\Omega} (C\partial_e u - u) < -\frac{1}{8n},$$

which contradicts (5.7); observe that we have assumed $r = 1$. Then for any $z \in \Gamma(u) \cap B_{1/2}$

$$u > 0 \quad \text{in } (z + \mathscr{C}_\delta) \cap B_{1/2}, \quad \text{and} \quad u = 0 \quad \text{in } (z - \mathscr{C}_\delta) \cap B_{1/2}.$$

In particular, $\Gamma(u) \cap B_{1/2}$ is a Lipschitz graph $x_n = f(x')$ with the Lipschitz constant of f not exceeding δ (Fig. 5.5).

This, in the original scaling, implies that there exists $\rho = \rho(u) > 0$ and a Lipschitz function $f : B'_\rho \to \mathbb{R}$ such that

$$\Omega_+ \cap B_\rho = \{x \in B_\rho : x_n > f(x')\}, \quad \text{and}$$
$$\Gamma_+ \cap B_\rho = \{x \in B_\rho : x_n = f(x')\}.$$

Moreover, if for $r_\delta, \delta \in (0, 1]$ is small enough such that we have

$$\|u_{r_\delta} - u_0\|_{C^{1,\alpha}(B_1)} \leq \epsilon, \quad \text{where } u_0(x) = \frac{1}{2}(x_n^+)^2,$$

Fig. 5.5 Cone of monotonicity $\mathscr{C}_\delta$ and Lipschitz regularity of the free boundary

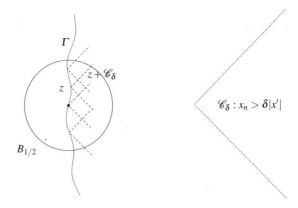

then $|\nabla_{x'} f| \leq \delta$ a.e. on $B'_{r_\delta/2}$. Since $\delta > 0$ is arbitrarily, we conclude the existence of a tangent plane to FB at the origin. This can be done similarly for points close to the origin, and hence we obtain C^1-graph, locally.

5.4 Other Types of FBPs

5.4.1 Bernoulli Type FB

Consider the functional

$$J(u) = \int_{B_1} \frac{1}{2}|\nabla u|^2 + \lambda_+^2 \chi_{\{u>0\}} + \lambda_-^2 \chi_{\{u<0\}},$$

where $\lambda_\pm > 0$, B_1 is the unit ball, and we have some fixed (but non-important) boundary data. For clarity of exposition we stick to the case $\lambda_+ = \lambda_- = 1$, i.e.

$$J(u) = \int_{B_1} \frac{1}{2}|\nabla u|^2 + \chi_{\{u \neq 0\}}. \tag{5.9}$$

Alt-Caffarelli-Friedman (ACF) considered this functional (TAMS '84), and proved existence and regularity for solutions. However, free boundary regularity was treated only for

$$J_1(u) = \int_{B_1} \frac{1}{2}|\nabla u|^2 + \lambda_+^2 \chi_{\{u>0\}} + \lambda_-^2 \chi_{\{u\leq 0\}}.$$

Observe the "tiny" difference between this and our functional. The minimizer u of J_1 is such that:

$$\text{interior}\{u = 0\} = \emptyset.$$

Fig. 5.6 The typical picture
of a Branch-Point

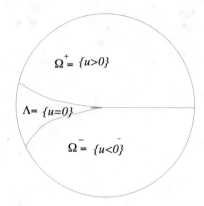

$\Omega^+ = \{u>0\}$

$\Lambda = \{u=0\}$

$\Omega^- = \{u<0\}$

The above functional introduces a new type of free boundary point, called *branch points*. The functional has a minimizer by lower semi-continuity, and the minimizer is Lipschitz regular. This can be found in the work of ACF. The regularity of the free boundary is, however, a different issue! I shall focus on the regularity of the free boundary at branch points, and give a fast track to the proof. For a more detailed account of this problem see [2] (Fig. 5.6).

As an application consider a huge (two-dimensional) block of a mushy-region D (mixture of ice and water) where the mean temperature is zero. Two disjoint sources of temperatures are placed in this mushy region $D_\pm \subset D$, with temperatures ± 1 on $D_\pm$ (one heating and one cooling). The temperature sources are ignited at time zero, and as time goes, both heat sources distribute themselves. The region D_+ naturally turns to water and expands, and in D_- freezes to ice and expands. As time grows large, we may see a stationary phase transition problem, with three distinct phases. The problem thus described (with three phases) maybe describe as follows

$$\Omega^+ := \{u > 0\}, \qquad \Omega^- := \{u < 0\}, \qquad \Omega^0 = \{u = 0\},$$

with u being temperature. Also along the FB $\partial\{\pm u > 0\}$ there should be some governing law. We shall consider the stationary case of the problem, and discuss the free boundary close to the so-called branch points, i.e. points where all three phases are non-trivially present.

Let us also define

- Free boundary: $\Gamma_u = (\partial\Omega^+ \cup \partial\Omega^-) \cap B_1$.
- One-phase: $\Gamma' = \{x \in \Gamma;\ \exists\, r > 0 :\ \overline{\Omega^+} \cap \overline{\Omega^-} \cap B_r(x) = \emptyset\}$.
- Two-phase : $\Gamma'' = \{x \in \Gamma :\ \forall\, r > 0 \text{ we have } \overline{\Omega^+} \cap \overline{\Omega^-} \cap B_r(x) \neq \emptyset\}$.
- Branch points: $\Gamma^* = \overline{\Gamma'} \cap \overline{\Gamma''}$.

We expect the following boundary condition

$$\frac{\partial u}{\partial \nu_+} = -\sqrt{\left(\frac{\partial u}{\partial \nu_-}\right)^2 + \lambda_+^2 - \lambda_-^2} \quad \text{on } \Gamma''$$

$$\frac{\partial u}{\partial \nu_\pm} = \mp\lambda_\pm \qquad\qquad\qquad \text{on } \Gamma'_\pm,$$

where by the subscripts "$\pm$" we indicate whether the outward normal ν is considered with respect to Ω^+ or Ω^-. This can be shown in a weak sense by variations of the form $u_\epsilon(x) = u(x + \epsilon\eta(x))$, and then looking at

$$\lim_{\epsilon \to 0} \frac{J(u_\epsilon) - J(u)}{\epsilon} = 0$$

Since we are dealing with the case $\lambda_+ = \lambda_- = 1$, we should formally have

$$\frac{\partial u}{\partial \nu_+} = \frac{\partial u}{\partial \nu_-} \text{ on } \Gamma'' \quad \text{and} \quad \frac{\partial u}{\partial \nu_\pm} = \mp 1 \text{ on } \Gamma'_\pm.$$

Next, we define two-plane solutions, denoted TP, those local minimizers u such that (in a rotated and translated coordinate system)

$$a\left(x_n^+ - (x_n + c)^-\right), \quad (a \geq 1, \ c \geq 0).$$

For branch points, or two-phase points we have $c = 0$, and we have a linear function ax_n, and also blow-ups are TP-solutions

$$\lim_{r \to 0} \frac{u(rx + x^0)}{r} \in TP.$$

We have the following result.

Theorem 5.2 ([2]) *Let u be a minimizer of $J(.)$. Then there exists a universal constant $\delta_0 > 0$ such that if*

$$\inf_{q \in TP} \|u - q\|_{\tilde{L}^2(B_1)} \leq \delta_0$$

then $\partial\Omega^\pm \cap B_{1/2}$ consists of two C^1-graphs

$$\{(x', f^\pm(x')); \ |x'| < 1/2\}.$$

Here $\tilde{L}^2$ is the average L^2-space.[17] A further, and more delicate, result is the structure of the branch points, formulated in the next theorem.

Theorem 5.3 ([2]) *There exist universal constants $\delta_0 > 0$ such that if u is a minimizer of J, $0 \in \Gamma_u^* \cap B_{1/2}$ and that the following conditions are satisfied:*

$$\|u - x_n\|_{\tilde{L}^2(B_1^+)} =: \delta \leq \delta_0, \tag{5.10}$$

$$\|\nabla'' u\|_{\tilde{L}^2(\Omega^+ \cup \Omega^-)} \leq \delta_0\delta, \tag{5.11}$$

---

[17]We use the notation $\tilde{L}^2$ mostly for simplicity of keeping tracks of scaling in the text.

$$\left\| u - c p_{3/2} \right\|_{\tilde{L}^2(\Omega^+ \cup \Omega^-)} \le \delta_0 \delta, \tag{5.12}$$

then the set of branch points $\Gamma^* \cap B_{1/4}$ is contained in an $(n-2)$–dimensional C^1 manifold.

Here $\nabla'' = (0, \partial_2, \dots, \partial_{n-1}, 0)$, and the above conditions mean:

- Condition (5.10) means standard $\tilde{L}^2$-closeness to a linear solution,
- Condition (5.11) means the derivatives in $x_2, \dots, x_n$ directions become appropriately small.
- Condition (5.12) means close to the first solution to the Thin obstacle problem. It is a kind of non-degeneracy.

Definition (Flatness) We say u is δ-flat in $B_r(x^0)$ if

$$\inf_{q \in TP} \| u - q \|_{\tilde{L}^2(B_r(x^0))} \le \delta.$$

Denote this by $u \in F(\delta, r, x^0)$. We now sketch the proof for Theorem 5.2.
Step 1: For each $\epsilon > 0$ there is a $\rho_\epsilon > 0$, and $\delta_0 > 0$ (universal), such that

$$\delta < \delta_0 \text{ and } u \in F(\delta, 1, 0) \quad \Rightarrow \quad u \in F(\epsilon \rho_\epsilon \delta, \rho_\epsilon, 0)$$

Moreover

$$\| q_1 - q_{\rho_\epsilon} \|_{\tilde{L}^2(B_1)} \le \delta,$$

where q_1 and q_{ρ_ϵ} are the corresponding TP-planes in the flatness. Here ρ_ϵ depends on the regularity of the $3/2$-homogenous solution $p_{3/2}(r, \phi) = r^{3/2} \cos(3\phi/2)$ of the thin obstacle problem.
Step 2: Flatness propagates to neighboring points:

$$u \in F(\delta, 1, 0) \quad \Rightarrow \quad u \in F(2^n \delta, 1, x^0) \quad \text{for all } x^0 \in B_{1/2}(0).$$

Step 3: Invert the relation $\epsilon \to \rho_\epsilon$, to obtain $\epsilon(\rho)$, as modulus of continuity, for smoothness:

$$u \in F(\delta, 1, 0) \quad \Rightarrow \quad u \in F(\rho \epsilon_\rho \delta, \rho, 0)$$

and control of the measure-theoretic normal vector

$$\| q_{\rho_1} - q_{\rho_2} \|_{\tilde{L}^2(B_1)} \le \epsilon(|\rho_2 - \rho_1|) \delta.$$

The above means that the linear polynomials q_ρ (i.e. the measure theoretic normals) converge (with ϵ modulus), and this gives that the free boundary is C^1.

We shall only give a hand-waving discussion of the proof of Step 1. The other steps are treated in standard ways, as done in minimal surface theory, and other free boundary problems, and hence omitted. If Step 1 does not hold then there exist $\delta_j \searrow 0$, $u^j \in F(\delta_j, 1, 0)$ and

$$u^j \notin F(\epsilon \rho_\epsilon \delta_j, \rho_\epsilon, 0). \tag{5.13}$$

Since in a rotated system $u^j(x) \to ax_n$, we need to consider the linearized function

$$v^j(x) := \frac{u^j(x) - ax_n}{\delta_j} \chi_{\Omega_j^+ \cup \Omega_j^-} \quad \to \quad what?$$

This is the heart of the matter! The most technical part is to show that there exists a universal C:

$$\|\nabla v^j\|_{L^2(\Omega_j^\pm \cap B_{3/4})} \le C \|u^j\|_{L^2(B_1)}, \ \forall \ j.$$

Once we do this, then we are able to find the limit of v^j. Here, We shall not discuss the convergence proof, but focus on what kind of equation the limit will satisfy. So for this we consider a sub-sequence of v^j that converges:

- $v^j \to w$ in $L^p(B_{3/4})$ for each $p < \infty$,
- $\nabla v^j \chi_{\Omega_j^+ \cup \Omega_j^-} \to \nabla w$ strongly in $L^q(B_{3/4})$ for all $q < 2$,
- $\Delta w = 0$ in $B_1(0) \setminus \Pi$ ($\Pi = \{x_n = 0\}$),
- By domain variation it follows that, for $\eta \in C_0^1(B_1 \cap \Pi)$,
-

$$\int_{B_1 \cap \Pi} \frac{\partial w^-}{\partial x_n} \eta = \int_{B_1 \cap \Pi} \frac{\partial w^+}{\partial x_n} \eta.$$

Here $w^\pm$ indicates that the derivative is taken on the upper or lower side of Π.

Moreover for $w^\pm(x') = \lim_{x_n \to 0^\pm} w(x', x_n)$ we have $w^+(x') - w^-(x') \le 0$, and $\frac{\partial w^-}{\partial x_n} = \frac{\partial w^+}{\partial x_n} = 0$ in the set $\{x \in \Pi; \ w < 0\}$. This implies w solves the well-known Thin Obstacle problem.

Linearization at other free boundary points leads to different limiting problems:

- Linearization at One-Phase free boundaries leads to a Neumann boundary condition on Π.
- Linearization at (non-branching) Two-Phase free boundaries leads to a (matching) Cauchy problem.

- It is well-known that solutions to the Thin Obstacle problem are universally $C^{1,1/2}$ in $\overline{B_{1/2}^\pm}$.
- Solutions to Neumann problems are C^2 in $\overline{B_{1/2}^\pm}$.
- Solutions to Matching Cauchy problem are C^2 in $\overline{B_{1/2}^\pm}$.

Conclusion There exists linear polynomials q such that

$$(\lim v^j =:) \quad w \approx q \text{ in } C^{1,1/2} - \text{norm.}$$

Since $w(0) = \nabla w(0) = 0$ (for the Thin obstacle case) we can take (for simplicity) $q \equiv 0$

$$\|w\|_{L^p(B_\rho)} \le \frac{1}{2}\rho^{1+\alpha} \quad \forall \, \alpha < 1/2, \ \rho \text{ small enough.}$$

Next going back to (5.13) we have

$$\inf_{q \in TP} \|u^j - q\|_{\tilde{L}^2(B_{\rho_\epsilon})} \ge \epsilon \rho_\epsilon \delta_j.$$

Define $\tilde{u}^j = \frac{u^j - q_j}{\delta_j}$, where q_j is the linear polynomial that is closest to u^j in $\tilde{L}^2(B_1)$, and recall that there is a small difference between $\tilde{u}^j$ and v^j (the linearization sequence): $\tilde{u}^j = v^j + \frac{q_j \chi_{A_j}}{\delta_j}$ where $A_j = B_1 \setminus (\Omega_j^+ \cup \Omega_j^-)$, with the property that $|A_j| = o(\delta_j)$. Now the contradictory argument boils down to

$$\inf_{q \in TP} \|v^j + \frac{q_j \chi_{A_j}}{\delta_j}\|_{\tilde{L}^2(B_{\rho_\epsilon})} \ge \epsilon \rho_\epsilon.$$

Recalling that v^j converges to w (and the worst regularity for w is $\rho_\epsilon^{3/2}$) we conclude

$$\|v^j - w\| \ge \epsilon \rho_\epsilon - \|\frac{q_j \chi_{A_j}}{\delta_j}\|_{\tilde{L}^2(B_{\rho_\epsilon})} - \rho_\epsilon^{3/2}.$$

As $j \to \infty$

$$0 \leftarrow \|\tilde{v}^j - w\|_{\tilde{L}^2(B_\rho)} \ge \epsilon \rho_\epsilon - o(\delta_j) - \rho_\epsilon^{3/2}.$$

This is a contradiction if we choose $2\rho_\epsilon < \epsilon^2$, and Step 1 follows.

5.4.2 Broken PDEs with FB

Our second problem here relates to FBs arising from standard PDEs where the ingredients are non-smooth.

The following equations $u'' = 0$, or/and $v'' - 1 = 0$, are solved by $u = ax + b$, and $v = \frac{x^2}{2} + ax + b$. The above equations produce smooth solutions because the ODEs are smooth. On the other hand the functions

$$h_1(x) = \begin{cases} x, & x < 0 \\ 2x, & x \geq 0 \end{cases} \qquad h_2(x) = \begin{cases} 0, & x < 0 \\ \frac{x^2}{2}, & x \geq 0 \end{cases}$$

are not smooth and cannot solve any smooth ODE. If we now define non-smooth (BROKEN) operators

$$L_1 w := \left((1 + \chi_{\{w>0\}})w'\right)', \qquad L_2 w := w'' - \chi_{\{w>0\}},$$

then h_1, h_2, above will solve (weakly) the ODEs, $L_1 h_1 = 0$, and $L_2 h_2 = 0$. Since $w^0 = \chi_{\{w>0\}}$, one may interpret $w^0 = \chi_{\{w>0\}}$, as a particular case of $(w^+)^a$ (with $a = 0$).

With these examples in mind, we shall now consider more complicated situations. I shall now give motivational examples in material sciences, and composite materials. Similar examples can probably be found in other applied areas, whenever there is a diffusion and phase changes/transitions. Consider a composite (say the unit ball B_1) made of two different materials M_1, M_1. The aim is to optimize physical quantities such as

- Material conductivity (of heat or electrical current)
- Resonance, vibration (first eigenvalue)

Now if D_i represents the region of material M_i and $B_1 = D_1 \cup D_2$ then one can write the corresponding PDEs in each cases above. The conductivity problem will be represented either by minimizing a functional of the type

$$\int_{B_1} a_1 \chi_{D_1} |\nabla u|^2 + a_2 \chi_{D_2} |\nabla u|^2 + other\ terms$$

or by the PDE

$$\operatorname{div}\left((a(x)\chi_{D_1} + b(x)\chi_{D_2})\nabla u\right) = R.H.S.$$

The resonance problem can be represented by

$$\operatorname{div}(A(x)\nabla u) = (a\chi_{D_1} + b\chi_{D_2} - \lambda)u.$$

Here $a, b, \ldots$ are given (functions) by the problem, and λ the first eigenvalue. One can simplify these (with new $a, b, A, ..$) as

$$\int_{B_1} (a\chi_D + b)|\nabla u|^2 + other\ terms$$

$$\mathrm{div}\,((a + b\chi_D)\,\nabla u) = R.H.S.$$

$$\mathrm{div}\,(A(x)\nabla u) = (a\chi_D u - \lambda(\alpha))u + b.$$

Remark 5.4 (Weighted Sobolev/Poincaré) A question that might be of interest is the weighted Sobolev/Poincaré constant related to the Broken PDEs, In other words one may as for sharp constants

$$\int_{Domain} u^2 \leq C_0 \int_{Domain} (1 + b\chi_{\{u>t\}})|\nabla u|^2$$

which relates to the broken functional (5.14) in the above.[18]

5.4.2.1 Quasi-Linear Case: Relation to Composites

Suppose we are forced to use a certain amount of each material. Or in this new formulation $W(|D|) = \alpha$, with W being a given function. The simplest case is volume case,

$$W(|D|) = |D| = \alpha,$$

and we further have the freedom of relocating D anyway we want to, and we want to see what the optimization problem will be. Hence finding D is also part of the problem.

The theory of shape optimization leads us to

$$D = \chi_{\{u>t\}}, \qquad for\ some\ t \in \mathbb{R}.$$

So the problems can now be written as

$$\int_{B_1} (a\chi_{\{u>t\}} + b)|\nabla u|^2 + other\ terms \tag{5.14}$$

$$\mathrm{div}\left((a + b\chi_{\{u>t\}})\,\nabla u\right) = R.H.S. \tag{5.15}$$

$$\mathrm{div}\,(A(x)\nabla u) = (a\chi_{\{u>t\}} + b - \lambda(\alpha))u. \tag{5.16}$$

---

[18]This could be a promising area of research with many interesting questions, such as geometric features and regularity theory, to be posed.

The above conductivity problem (5.15) has been considered earlier by Andersson–Mikayelyan [1] for special matrices A, and the parabolic case by Caffarelli–Stefanelli [7], cf. also [18], see also [19].

I shall set $t = 0$ for simplicity, and discuss mostly Eqs. (5.15), and (5.16). One may play around with constraints of type

$$\int_D (u^+)^a dx = \alpha$$

then we end up with other type of PDEs

$$\text{div} \left((a + b(u^+)^a) \nabla u \right) = R.H.S.$$

and the resonance problem by

$$\text{div} (A(x) \nabla u) = a(u^+)^a + b - \lambda u.$$

A further connection of the problem can be made towards transmission problem. These ideas from shape optimization and rearrangement techniques is an invitation to consider a new class of problems coming from broken PDEs. The well-known transmission problems, i.e. when D is fixed, is thus partially related to our problem, but has a different angle of difficulty. In the transmission problem, one assumes regularity of ∂D and looks for uniform regularity of solutions in each region. Here in our model the domain D is unknown.

5.4.2.2   Semilinear Case: Relation to Obstacle Type Problems

The resonance problem and similar PDEs with broken RHS, have been considered by several people (Chanillo-Kenig, Andersson-Sh. -Weiss, ...) The most general form of the above semilinear PDE can be written as

$$\Delta u = a_+(x)(u^+)^q - a_-(x)(u^-)^q.$$

The analysis bifurcates due to the sign of $a_+ + a_-$. For $a_+ + a_- < 0$ the problem has interesting surprises, and has been considered earlier by several authors (for the case $q = 0$). The case $a_+ + a_- > 0, 0 < q < 1$, is completely new and some partial results appear in [13], but many challenging questions are left untreated.

5.4.2.3   Local Analysis and Regularity

Here I shall discuss only the quasilinear case, see [18]. For semilinear problems see [13], and the references therein.

There are two questions (among many others) that do not follow by classical arguments:

- Regularity of any solution u.
- Regularity of the level "surface" (nodal sets) $\partial\{u > 0\}$.

Recall that the PDE breaks exactly on $\partial\{u > 0\}$. So classical theory fails across this set.

For the quasilinear problem (here I take $a = b = 1$)

$$div(1 + \chi_{\{u>0\}})\nabla u = f(x, u), \qquad f \quad \text{bounded}.$$

one can see that the function $v = u + u^+$ solves a semilinear eigenvalue problem

$$\Delta v = g(x, v) \qquad g, \quad \text{bounded}.$$

From the last equation we have that $|\Delta v| \leq C$, so that $v \in C^{1,\alpha}$ [19] and hence ∇v is universally bounded, i.e. $|\nabla u| \leq C_0$. This is the best regularity to expect from such problems.

Speculation is that Lipschitz regularity may fail, for general operators with matrix coefficient $div(I + A(x)\chi_{\{u>0\}})\nabla u)$. There were some convincing discussions with John Andersson, that points towards Lipschitz failure in general.

Theorem 5.5 *Let a and b be Dini continuous, $\lambda_0 < a$, $b < \lambda_1$, and u be a weak solution to*

$$div((a(x) + b(x)\chi_{\{u>0\}})\nabla u) = 0, \qquad in \ B_1.$$

Then u is Lipschitz in $B_{1/2}$, with a universal Lipschitz norm.

Next define

$$\Gamma(u) = \partial\{u > 0\}, \qquad \mu := div((a + b)\nabla u^+)$$

then the following theorem holds.

Theorem 5.6 (Smoothness of Non-degenerate Nodal Sets) *Let a and b be α-Hölder continuous, and u be a weak solution to our problem in B_1. Then for μ-a.e. $z \in \Gamma(u)$, there is $r > 0$ such that*

$$\Gamma(u) \cap B_r(z) \quad is \ a \ C^{1,\alpha} \ graph.$$

Here the radius r and the $C^{1,\alpha}$ norm depend on u, z and the Hölder norm of a, b. (Compare with nodal sets of harmonic function $h = xy$.)

---

[19]Observe that $v \in C^{1,\alpha}$ does not help to obtain regularity results for u since $v = u + u^+$.

5.4.2.4 Proof of Lipschitz Regularity

By DeGiorgi, we have solutions are C^β for some $\beta > 0$.

Given this initial regularity, one can actually bootstrap easily to the case of any $\beta < 1$. A simple blow-up argument will do that.

To prove that "little" left to be Lipschitz, we need a strong machinery (essentially due to Alt-Caffarelli-Friedman), which claims that the r-dependent function

$$\Phi(r, z, h_1, h_2) := \frac{1}{r^4} \int_{B_r(z)} \frac{|\nabla h_1|^2}{|x|^{n-2}} dx \int_{B_r(z)} \frac{|\nabla h_2|^2}{|x|^{n-2}} dx, \qquad (5.17)$$

is monotone in r, whenever $h_1 h_2 = 0$, $h_1(z) = h_2(z) = 0$ they are subharmonic, non-negative, and in $W^{1,2}$. The monotonicity-argument above has been extended to various situations, but always with same divergence type PDE on both sides, for h_1, and h_2.

Proposition *There is a universal constant C, depending on the $W^{1,2}$ norm of u such that our solution u satisfies*

$$\Phi(r, z, u^+, u^-) \leq C, \qquad \forall z \in \Gamma(u) \cap B_{1/2}.$$

Suppose this is proven, and suppose that ∇u kind of exists at z (at least in the sense of Lebesgue point), then by Fatou's lemma

$$c_0 |\nabla u^+(z)|^2 |\nabla u^-(z)|^2 \leq \Phi(r, z, u^+, u^-) \leq C, \qquad \forall z \in \Gamma(u) \cap B_{1/2}.$$

Hence (formally) if one of them is infinity, the other must be zero. This gives us a good ground for proving that both have to be bounded. The proof uses blow-up technique and contradictory argument, along with this monotonicity function.

5.4.2.5 Proof of Smoothness of Nodal Sets

To prove regularity of the Nodal sets $\Gamma(u)$ we take these steps:

(a) Degenerate points have μ-measure zero. Hence non-degenerate points have full μ-measure.
(b) For μ-a.e. $z \in \Gamma(u)$, there exists a function

$$u(x) \approx \beta_+(x \cdot v)^+ - \beta_-(x \cdot v)^- \quad \text{in } B_{r_z}(z),$$

with $\beta_\pm$ given by the ingredients.
(c) Reiterate (b) to show that the asymptotic representation holds for each $z' \in \Gamma(u) B_{r_z}(z)$.
(d) Show u is a viscosity solution to the Bernoulli FB, and use standard techniques to prove $C^{1,\alpha}$.

The ideas originate in the work of Andersson-Mikayelyan, combined with seminal work of L. Caffarelli on regularity of FB problems of Bernoulli type.

Remark 5.7 A recent result by the author, Kim, and Lee [19] presents a more effective proof of the above results as well as introduces several further results. The method in [19] is a freezing point technique and Schauder estimates.

5.4.3 Non-variational Problems

There are many problems that can not be represented as minimization or variational inequalities. One of them that we encountered was the broken-PDE

$$\text{div}((a + b\chi_{\{u>0\}})\nabla u) = 0.$$

Another example is the well-studied no-sign obstacle problem[20]

$$\Delta u = \chi_{\{u \neq 0\}},$$

or even more general

$$\Delta u = \chi_{\{|D^2 u| \neq 0\}},$$

that arise in applications in potential theory, and the theory of quadrature domains; see [8].

Regularity results for solutions of such problems are usually much harder and does not work through standard techniques of free boundaries. In these cases one is forced to use stronger tools such ACF monotonicity function. However, once the operator changes from Laplacian to others then the chances of applying ACF diminished drastically.

The full strength of ACF monotonicity function, (5.17) is realised when we consider applications to problems of this kind, or more generally to semilinear PDEs $\Delta u = f(x, u)$ with f Lipschitz in x and $f'_u \geq -C$. This seems to be a borderline for applying this

We shall next derive optimal regularity of the non-variational problem, through techniques that are very robust and can be applied to many other problems, such fully nonlinear cases.

The nonlinear theory, i.e. replacing the Laplacian with $F(D^2 u)$ has also been done recently [11] as well as parabolic versions of these [12]. The p-Laplacian operator, Monge-Ampere, and many other operators, are yet not treated. One may also look at potentials of fractional order.

---

[20]If we try to represent this as Euler Lagrange equation of a minimizer, we only end up with solutions to $\Delta v = 1$.

Another variant that one may consider is

$$\Delta u_p = |u_p|^p$$

and what happens when $p \to 0$.

I shall put this problem in terms of potential theory and the optimal regularity of potentials. This is more general and can be adapted to different situations. Let U^D denote the potential of a (bounded) domain D in $\mathbb{R}^n$

$$U^D(x) = c_n \int_D |x - y|^{2-n} dy \qquad (n \geq 3)$$

and for $n = 2$ we have the logarithmic potential. Here c_n is a normalization factor. In general we shall consider smooth densities f and the weighted potential

$$U^{D,f}(x) = c_n \int_D \frac{f(y)dy}{|x - y|^{n-2}} dy.$$

5.4.3.1   Regularity of Potentials (Heuristics)

For simplicity of notation we shall always write U for the potentials with density $f \chi_D$. It is well known that a potential U satisfies (in the distributional sense)

$$\Delta U = f \chi_D.$$

In particular U has bounded Laplacian, and one obtains from well-known classical Schauder theory that $U \in C^{1,\alpha}$ for any $\alpha < 1$. Note that U is not C^2 across the boundary of D. Let us assume $D^2 U$ is uniformly bounded outside D. Since U is $W^{2,p}(B_1)$ $(p > n)$ then it is C^2 almost everywhere in B_1.

We want to show $D^2 U(x)$ is universally bounded in $B_{1/2}$. Fix any point $z \in B_{1/2}$ and set

$$\bar{U} := U(x) - U(z) - (x - z) \cdot \nabla U(z).$$

$$S_r = \sup_{B_r(z)} |\bar{U}(x)|/r^2.$$

We need to show that for a universal C_0

$$S_r \leq C_0 \qquad \forall r < 1/2.$$

Define

$$\lambda_r = |B_1(0) \setminus D_r|, \qquad \text{where } D_r := \frac{1}{r}(D - z).$$

Proposition 5.8 (J. Andersson) *There is a universal constant M such that for any $0 < r < 1$ either of the following hold*

- $S_r \leq M,$ *this is what we want*
- $\lambda_r \leq \frac{1}{2}\lambda_{2r},$ *this says the complement is polynomially thin.*

We assume, for the moment, this true. Then the first observation is that

$$\lambda_r \leq \frac{1}{2}\lambda_{2r} \quad \Longrightarrow \quad \lambda_r \lesssim r,$$

i.e.

$$|D^c \cap B_r| \leq C r^{n+1}.$$

We thus arrive at

$$D^2 U^{D^c}(z) \approx \int_{D^c \cap B_1(z)} \frac{dy}{|y - z|^n} \leq C.$$

Let us see how we use this estimate. Set $r_k = 2^{-k}$, and consider two cases:

- $\liminf_k S_{r_k} \leq 3M,$
- $\liminf_k S_{r_k} > 3M.$

In the first case we obtain (recalling first that $\bar{U} := U(x) - U(z) - (x - z) \cdot \nabla U(z)$)

$$|D^2 U(z)| = |D^2 \bar{U}(0)| \leq \liminf_{k \to \infty} \sup_{B_{2^{-k}}(0)} \frac{2|\bar{U}|}{2^{-2k}} \leq 2(C_1 + 3M).$$

In the second case, since $S_1 \leq 3M$, there exists k_0 such that:

$$S_{r_{k_0}} \leq 3M, \quad \text{and} \quad S_{r_k} > 3M, \qquad \forall k \geq k_0.$$

In particular by Andersson's proposition one has

$$\lambda_r \leq Cr \qquad \forall r \leq 2^{-k_0}.$$

Now set $U_{r_{k_0}}(x) := 2^{-2k_0} \bar{U}(2^{-k_0}x + z)$. Then

$$\left| \bar{U}_{r_{k_0}} \right|(x) \leq 3M \qquad \text{in } B_1(0).$$

We may write

$$\bar{U}_{r_{k_0}}(x) = w(x) - \bar{U}^{D^c_{r_{k_0}}}(x),$$

where now $\Delta w = f(2^{-k_0}x + z)$, $|w| \le C$ in B_1, and f is Dini. In particular $|D^2 w(0)| \le C$. From here we arrive at

$$D^2 U(z) = D^2 \bar{U}_{r_{k_0}}(0) = D^2 w(0) + D^2 \bar{U}^{D^c_{r_{k_0}}}(0).$$

It remains to prove bound for the last term, which can be rewritten in terms of an integral, and we have

$$|D^2 \bar{U}^{D^c_{r_{k_0}}}(0)| \le \int_{D^c_{r_{k_0}} \cap B_1} \frac{dx}{|x|^n} \le \int_{r_1}^1 \frac{\chi_{D^c_r} dr}{r} \le C,$$

for $r_1 = dist(z, \partial D_{r_{k_0}})$. Here we have used

$$\lambda_r \le Cr \qquad \forall r \le 2^{-k_0}.$$

This gives the result.

5.4.3.2   Proof of John Andersson's Dichotomy

Set $\bar{U}_r(x) = \bar{U}(rx)/r^2$, and let v_r be such that

$$\Delta v_r = -f(rx)\chi_{B_1 \setminus D_r}, \qquad v_r = 0 \text{ on } \partial B_1.$$

Then $\bar{U}_r = w_r + v_r$, with $\Delta w_r = f(rx)$ and w_r has the information of supnorm of $\bar{U}_r$ on ∂B_1. Also

$$\int_{B_{1/2}} |D^2 v_r|^2 \le C|D^c_r \cap B_1| = C\lambda_r.$$

For clarity we assume $D^2 \bar{U} = 0$ in D^c. Next

$$0 = \int_{D^c_r \cap B_{1/2}} |D^2 \bar{U}_r|^2 = \int_{D^c_r \cap B_{1/2}} |D^2 w_r + v_r|^2.$$

In particular (by triangle inequality and the above analysis)

$$\int_{D^c_r \cap B_{1/2}} |D^2 w_r|^2 \le \int_{D^c_r \cap B_{1/2}} |D^2 v_r|^2 \le C\lambda_r.$$

Let now $\tilde{w}_r = w_r/S_r$, then (for $S_r \ge M$ large) we have

$$\int_{D^c_r \cap B_{1/2}} |D^2 \tilde{w}_r|^2 \le \frac{C}{S_r^2}\lambda_r \le \frac{C}{M^2}\lambda_r,$$

with $\tilde{w}_r$ solving $\Delta \tilde{w}_r = f(rx)/S_r$ and $\sup_{B_1} |\tilde{w}_r| = 1$. Now we need

$$C_1 \lambda_{r/2} \leq \int_{D_r^c \cap B_{1/2}} |D^2 \tilde{w}_r|^2 \leq \frac{C}{M^2} \lambda_r,$$

which would give $2\lambda_{r/2} \leq \lambda_r$, if M is large enough.

This is another tricky part!

We need a kind of non-degeneracy for $|D^2 \tilde{w}_r|^2$ on the set $D_r^c \cap B_{1/2}$. Here is how we do it.

For the first inequality above we may now split $\tilde{w}_r$ into two parts:

$$\tilde{w}_r = h_r + g_r$$

where h_r is homogeneous harmonic polynomial of degree two and g_r satisfies

$$\Delta g_r = f(rx)/S_r \quad \text{and} \quad g_r = 0 \text{ on } \partial B_1.$$

In this way we get rid of g_r as it becomes uniformly C^2, since f is Dini (say).

For h_r we have $D^2 h_r$ is a constant matrix, so we obtain the volume

$$c_0 \lambda_{r/2} = \int_{D_r^c \cap B_{1/2}} |D^2 h_r|^2$$

All to all we have

$$c_0 \lambda_{r/2} - c_1 \lambda_{r/2}/M = \int_{D_r^c \cap B_{1/2}} |D^2 h_r|^2 - \int_{D_r^c \cap B_{1/2}} |D^2 g_r|^2$$

$$\leq \int_{D_r^c \cap B_{1/2}} |D^2 \tilde{w}_r|^2 \leq \frac{C}{M^2} \lambda_r.$$

For M large enough we have

$$2\lambda_{r/2} \leq \lambda_r.$$

5.4.4 Nonlocal Problems, Extensions and Thin Obstacles

5.4.4.1 Thin Obstacles: Semipermeable Membranes

A semipermeable membrane is a membrane that is permeable only for a certain type of molecules (solvents) and blocks other molecules (solutes). Because of the chemical imbalance, the solvent flows through the membrane from the region of smaller concentration of solute to the region of higher concentration osmotic pressure. The flow occurs in one direction and continues until a sufficient pressure builds up on the other side of the membrane (to compensate for osmotic pressure), which then shuts the flow. This process is known as osmosis.

Let $D \subset \mathbb{R}^{n+1}$ be the region occupied with a chemical solution, $\mathcal{M} \subset \partial D$ be the semipermeable part of the boundary, and u be the pressure of the chemical (compressible) solution, that in the stationary case satisfies

$$\Delta u = 0 \quad \text{in } D.$$

We remark that we do not make any difference between the notation for the Laplacian in $\mathbb{R}^n$ or in $\mathbb{R}^{n+1}$.

Let $\psi : \mathcal{M} \to \mathbb{R}$ represents the (given) osmotic pressure. Then on $\mathcal{M}$ we have the following boundary conditions

$$u > \psi \quad \Rightarrow \quad \partial_\nu u = 0 \qquad \text{(no flow),}$$
$$u \leq \psi \quad \Rightarrow \quad \partial_\nu u = \lambda(u - \psi) \qquad \text{(flow),}$$

where λ is the permeability constant (finite permeability). Letting $\lambda \to \infty$ (infinite permeability), we obtain the Signorini boundary conditions on $\mathcal{M}$

$$u \geq \psi, \quad \partial_\nu u \geq 0, \quad (u - \psi)\partial_\nu u = 0.$$

This problem has attracted a lot of attention in the last decade, and the topic has advanced greatly. A more general notion of thin obstacle has been introduced through extension problems for the so-called fractional problems, that we discuss in the next section.

5.4.4.2 Nonlocal Problems

Motivated by modeling the American options for stocks with possible discontinuities, we consider a version of the optimal stopping problem for jump processes. Let $\mathbf{X}_t$ be an α-stable Lévy process, $(0 < \alpha < 2)$ with $\mathbf{X}_0 = x$ that will model the logarithm of the stock price. Let $\psi : \mathbb{R}^n \to \mathbb{R}$ be a payoff function in the sense that we generate a profit of $\psi(y)$ when trading the stock at $\mathbf{X}_t = y$. We then want to maximize the expected profit

$$u(x) = \sup_\theta \mathbb{E}\left[\psi(\mathbf{X}_\theta), \right] \tag{5.18}$$

where θ ranges over admissible stopping times.

If we define the derivative (or infinitesimal generator)

$$(-\Delta)^{\alpha/2} u := \lim_{t \to 0^+} \frac{1}{t} \mathbb{E}^x \left[u(x) - u(x + X_t) \right]$$

one can show that the value function u (in (5.18)) solves an obstacle-type problem

$$u \geq \psi, \quad (-\Delta)^{\alpha/2}u \geq 0, \quad (u - \psi)((-\Delta)^{\alpha/2}u) = 0 \quad \text{in } \mathbb{R}^n.$$

The above nonlocal integro-differential operator can also be defined[21] by the Fourier multiplier as

$$\widehat{(-\Delta)^{\alpha/2}u} = |\xi|^\alpha \widehat{u}(\xi).$$

In the case when $\alpha = 1$, this has a direct connection to the thin obstacle problem. Indeed, by adding an extra dimension, and extending u harmonically to $\mathbb{R}^{n+1}_+ = \mathbb{R}^n \times (0, \infty)$ by solving the Cauchy problem

$$\Delta \tilde{u} = 0 \quad \text{in } \mathbb{R}^{n+1}_+, \quad \tilde{u}(\cdot, 0) = u,$$

it is then easy to verify that $-\partial_{x_{n+1}}\tilde{u}(x, 0) = C_n(-\Delta)^{1/2}u(x)$ and thus $\tilde{u}$ solves a boundary thin obstacle problem in $\mathbb{R}^{n+1}_+$. While the problem for u is nonlocal, one effectively localizes the problem by adding an extra dimension.

A different interpretation of this problem is the consideration of the Dirichlet to Neumann operator (in $\{x_n > 0\}$) with u harmonic

$$T: \quad u_0 := u(x', 0) = u(x) \llcorner_{\mathscr{M}} \quad \rightarrow \quad -\partial_\nu u(x) \llcorner_{\mathscr{M}}$$

where $\mathscr{M}$ is the hyperplane $\{x_{n+1} = 0\}$. This operator is non-negative

$$(Tu_0, u_0) = \int_{\mathscr{M}} -\partial_\nu u(x)u_0 = \ldots = \int_{\mathbb{R}^n} |\nabla u|^2 \geq 0,$$

---

[21] There are many other ways for defining this operator, as can be found in the literatures. The fractional Laplacian through Fourier transform is defined as

$$\mathscr{F}[(-\Delta)^{\alpha/2}v](\xi) = |\xi|^{2s}\mathscr{F}[v](\xi)$$

that is

$$(-\Delta)^{\alpha/2}v(x) = \mathscr{F}^{-1}\left[|\xi|^\alpha \mathscr{F}[v]\right](x).$$

We also assume $0 < \alpha < 1/2$. Other values of s can be allowed as long as $|\xi|^\alpha$ is a tempered distribution. This amounts to $\alpha > -n$. The representation through Fourier transform leads to

$$(-\Delta)^{\alpha/2}v(x) = C_{n,\alpha}\text{P.V.} \int_{\mathbb{R}^n} \frac{v(x) - v(y)}{|x - y|^{n+\alpha}}dy,$$

which gives us a (non-PDE) tool to work with. Think of

$$H \star G = \mathscr{F}^{-1}(\mathscr{F}(H)\mathscr{F}(G)).$$

where we have assumed u tends to zero fast enough at infinity, and v is outward normal direction. This along with the fact that

$$T^2 u_0 = (-\partial_v)(-\partial_v)u(x) \lfloor_{\mathcal{M}} = -\Delta u(x) \lfloor_{\mathcal{M}},$$

where Δ is n-dimensional Laplacian on $\mathcal{M} = \{x_{n+1} = 0\}$, gives us

$$T u_0 = (-\Delta)^{1/2} u_0.$$

This suggests the connection between fractional laplacian on the lower-dimensional manifold $\mathcal{M} = \mathbb{R}^n$ to the extended problem in $\mathbb{R}^{n+1}$.

In particular, any study of the fractional problem can be translated into the extension problem, and studied in the large space.

5.4.4.3 Local Analysis of Fractional Obstacle Problem

The general case of $0 < \alpha < 2$ in the optimal pricing of American option results to different type of extension operators. We shall here discuss the corresponding extension of such problems.

It is an easy exercise to see that solutions to our obstacle problem are given by minimizing the functional

$$J(v) = \int_{\mathbb{R}^n} \int_{\mathbb{R}^n} \frac{|v(x) - v(y)|^2}{|x - y|^{n+2s}} dx dy,$$

over a suitable class with $v \geq \phi$. Just make a variation $v + \epsilon \psi$, with $\psi \geq 0$.
What is new here, and what is not easy?

- The operator is non-local, meaning: Changing the value of v at any point causes a complete change of v everywhere.
- Scaling of the "operator" does not justify the regularity. How does it scale?

The idea, that originated in the work of S. A. Molchanov, E. Ostrovskii, [21] was explored by Caffarelli-Silvestre to use extension theorems.

More exactly, one can see that the fractional Laplacian, as represented by integral identity

$$(-\Delta)^s v(x) = C_{n,s} \text{P.V.} \int_{\mathbb{R}^n} \frac{v(x) - v(y)}{|x - y|^{n+2s}} dy,$$

is somehow reminiscent of derivatives of Poisson representation of harmonic functions in $\mathbb{R}^{n+1}_+$. If we set

$$P(x - z, y) = C \frac{y^{1-a}}{(|x - z|^2 + y^2)}, \qquad (a = 2s - 1, \quad s = (a + 1)/2),$$

then

$$u(x, y) = \int_{\mathbb{R}^n} P(x - z, y)v(x)dz$$

solves the PDE

$$\text{div}(y^a \nabla u(x, y)) = 0 \quad \text{in } \mathbb{R}^{n+1}_+.$$

Furthermore, for the obstacle problem in its extension formulation, we have

$$u(x, 0) \geq \phi(x)$$

$$\lim_{y \to 0_+} y^a u_y(x, y) = 0 \qquad \text{for } u(x, 0) = v(x) > \phi(x)$$

and

$$\lim_{y \to 0_+} y^a u_y(x, y) \leq 0 \quad \text{in } \mathbb{R}^{n+1}.$$

In this formulation the problem is related to the so-called thin obstacle problem, where one takes $a = 0$ and reflects the solution on the plane $\{x_{n+1} = 0\}$ so that it is defined in $\mathbb{R}^{n+1}$.

The approach to solve the local problem, which also solves the non-local problem, follows a slightly different path than is common for FBP. The main reason is the regularity of solutions, that seems hard to capture, without classifying global solutions. Let us see how this works for $s = 1/2$ case.

The heart of the matter is optimal regularity of solutions. Once we have proved the optimal regularity, then the regularity of the FB follows more or less in the same spirit as that of the obstacle problem. A desired solution to our problem looks like (locally)

$$u(x) = C\text{Re}(x_1 + i|x_2|)^{3/2}, \qquad i = \sqrt{-1}.$$

An argument (essentially due to) Hans Lewy suggests considering $w = \partial_1 u \partial_2 u$, which is harmonic in the upper half ball and (formally) zero on $x_2 = 0$. By odd extension it becomes harmonic in the ball.

If P_κ is the first harmonic homogeneous polynomial (of order $\kappa \geq 1$) for the Taylor expansion of w, then for $w \approx r^\kappa$, we have $u \approx r^{1+\kappa/2}$. So for $\kappa = 1$ we have $u \approx r^{3/2}$ close to the origin. Observe that this argument does not seem to use the information $\partial_2 u \leq 0$.

- One first proves $C^{1,\alpha}$-regularity (this is classical).
- One shows that Almgren's Frequency function also holds for Thin obstacle problem $(Dirichlet_{B_r}/\|u(rx)\|_{L^2(\partial B_1)})$.

- One scales $u(rx)/\|u\|_{L^2(\partial B_r)}$, and obtains global homogenous solution.
- Global homogeneous solutions are classified, and the order κ of homogeneity are $\kappa = 1/2$, $2m$, or $2m - 1/2$, with $m \in \mathbb{N}$.
- Since we already have $C^{1,\alpha}$-regularity, we must have $\kappa \geq 3/2$.
- Prove optimal $C^{1,1/2}$-regularity, up to the set $\{x_{n+1} = 0\}$.

5.5 System Case

5.5.1 Optimal Switching

Evaluation of investment projects, in various applications (Political economics, Social sciences, Financial market) results in optimal switching problems. The problem in such applications is to make a decision under uncertainty, given several possible choices. Such decision may be any of

- Usage of gas, oil, or carbon fuel for heating (of a city).
- When to use which, and when is optimal to switch between them?
- Keeping a mine open or closed, depending on the market price of the commodity.

Let $S = S_t$ be the spot price of the commodity (Copper), and $F = F(S, t)$ the price of a Futures (right to deliver a commodity with specific price). The price S satisfies a stochastic differential equation

$$dS = \sigma S dW + r S dt,$$

where $dW^2 = dt$, r is interest rate, and σ market volatility for S.

The valuation of futures contracts F follows a similar reasoning as that in the option case. Hence F satisfies

$$F_t + \frac{1}{2}\sigma^2 S^2 F_{SS} + r S F_S - r F = 0,$$

Here we consider a European type Futures, so the exercise date is fixed. We also ignore various parameters, such as Convenience yield, and local trend in stock. From here one can easily deduce (Ito's formula) that

$$dF = \ldots = \sigma S F_S dW + (r S F_S + \frac{1}{2}\sigma^2 S^2 F_{SS} + F_t)dt$$

and inserting the PDE for F we have

$$dF = \sigma S F_S dW + r F dt.$$

Let $H = H_j = H(S, t, j, \psi)$ be the value of Copper Mine, where $j = 0, 1$ corresponds to open, respectively close state. Also set ψ to be mine's operating policy, and consider the portfolio

$$\Pi = H - mF$$

> Long position in Mine and short position in m Futures contracts.

Differentiating we obtain

$$d\Pi = dH - mdF = H_S dS + H_t dt + \frac{1}{2}\sigma^2 S^2 H_{SS} dt - m\sigma SF_S dW - mrFdt.$$

Now choosing

$$m = H_S/F_S, \qquad H_S dS = H_S(\sigma S dW + r S dt)$$

we have

$$d\Pi = (rSH_S + H_t + \frac{1}{2}\sigma^2 S^2 H_{SS})dt - mrFdt.$$

Also arbitrage theory tells us:

> Risk free Return $\geq$ Return from the portfolio

Hence we obtain the Black Scholes equation for H

$$r(H - mF)dt \geq (rSH_S + H_t + \frac{1}{2}\sigma^2 S^2 H_{SS})dt - mrFdt,$$

i.e.,

$$L(H) := H_t + rSH_S + \frac{1}{2}\sigma^2 S^2 H_{SS} - rH \leq 0,$$

where $H = H_j = H(S, t, j, \psi)$. We ignore the dependence on policy and set $H_j = H(S, t, j)$, for $j = 0, 1$ (open and closed state). Switching between states involves either:

> cost ψ_0 = rehiring, buying equipments

or

> profit ψ_1 = selling inventory.

In particular switching forth and back should be costly:

$$\psi_0 + \psi_1 > 0.$$

It is also notable that

$$H_0 \geq H_1 - \psi_0, \qquad H_1 \geq H_0 - \psi_1.$$

A final ingredient as in the case of option valuation is that H_j should be **minimum/smallest** among all such possible values.

Above equations lead to a system of variational inequalities

$$\min(L(H_i), H_i - H_j + \psi_i) = 0, \qquad i, j = 0, 1, \quad i \neq j,$$

with

$$\psi_0 + \psi_1 > 0,$$

and reasonable boundary and terminal/initial data. One may consider a system of m-equations, corresponding to m-different states.

The optimal switching problem has been treated from both stochastic- and PDE point of views where the general question has been existence and uniqueness of solutions. Regularity theory for solutions has been treated in special cases and there are several partial results. We refer to a recent survey on the topic [3] for an overview.

It is straightforward that if (H_1, H_2) is a (viscosity) solution to optimal switching problem then $H = H_1 - H_2$ solves the double-obstacle problem check to make it exact and correct

$$\min\{\max\{LH, H - \psi_2, H + \psi_1\}\} = 0. \tag{5.19}$$

From this rewriting it is apparent that one should expect similar type of results for the switching problem as that of double obstacle. Nevertheless, results that one may obtain in this formulation for H cannot be transferred to H_1, H_2, easily. E.g. the optimal regularity of H does not carry over to the original problem. On the other hand, since the free boundary for double-obstacle problem and optimal switching is the same, one may obtain regularity properties for the free boundary of the original problem by studying the optimal switching problem.

5.5.2 Minimization Problems

Let $f : \mathbb{R}^m \to R$ ($m \geq 1$) be bounded, and consider minimization of the functional

$$E(\mathbf{u}) := \int_{B_1} \left(|\nabla \mathbf{u}|^2 + f(\mathbf{u}) \right) dx.$$

Here B_1 is the unit ball in $\mathbb{R}^n$ ($n \geq 1$), and we minimize over $W_{\mathbf{g}}^{1,2}(B_1; \mathbb{R}^m)$, for some smooth $\mathbf{g} = (g_1, \cdots, g_m)$. The minimizer(s) are vector-valued functions $\mathbf{u} = (u_1, \cdots, u_m)$, and we call them energy minimizing maps.

I shall discuss two special cases of the function $f(\mathbf{u})$:

(1) $f(\mathbf{u}) = 2|\mathbf{u}|$.
(2) $f(\mathbf{u}) = \chi_{\{|\mathbf{u}|>0\}}$.

We assume further $\mathbf{u} = (u_1, \cdots, u_m)$, with $u_i \geq 0$, $\forall i = 1, \cdots, m$. The set $\{|\mathbf{u}| > 0\}$ competes with the Dirichlet energy. So contrary to standard PDE, and in general $\{\mathbf{u} = 0\} \neq \emptyset$. E.g. try with small boundary values, then it is better to pay more in Dirichlet energy than the volume $|\{|\mathbf{u}| > 0\}|$. The set $\partial\{|\mathbf{u}| > 0\}$ is called Free Boundary.

5.5.2.1   Case $f(\mathbf{u}) = 2|\mathbf{u}|$, $m = 2$: Reaction-Diffusion System

This is the equilibrium state of a cooperative reaction-diffusion system

$$u_t - \Delta u = -\frac{u}{\sqrt{u^2 + v^2}},$$

$$v_t - \Delta v = -\frac{v}{\sqrt{u^2 + v^2}},$$

describing the interaction between concentrations u and v of two species (reactants).

Each species (reactant) slows down the extinction (reaction) of the other species, and hence they cooperate.

5.5.2.2   Case $f(\mathbf{u}) = \chi_{\{|u|>0\}}$: Thermal Insulation

This case describes (optimal) stationary thermal insulation, allowing a prescribed heat loss from the insulating layer. Depending on the cost of insulation material, sharing of the insulation layer may become more optimal. The heat flows in from the boundary of the domain Ω, through a vector function $\mathbf{g} \in H^1(\Omega; \mathbb{R}^m)$ on the boundary (boundary data). Each g_i gives rise to a potential function u_i describing the heat distribution from the data g_i, and the system has to cost through Dirichlet energy as well as the total volume of heated region $|\{|\mathbf{u}| > 0\}|$. If the supports of g_i-s stay far from each other then it is reasonable that the system behaves exactly like scalar case, for each $i = 1, \cdots, m$. When the supports of g_i-s come close to each other (or g_i become large) then naturally the volume of each support $\{u_i > 0\}$ come closer, and at some stage it is less costly to use same insulation layer, i.e. they prefer to share support, and hence $\sup u_i = \sup u_j$ for some of these i, j. Those g_i that are still small will insulate separately. The total heat of the system at each point is given by $\sum u_i$, and this is a major difference between our problem and standard scalar problem.

5.5.2.3   Properties of Solutions

In both problems we have

$$\{u_i > 0\} = \{u_j > 0\}, \quad \text{for} \quad i, j = 1, 2 \cdots, m.$$

In the set $\{|\mathbf{u}| > 0\}$, we can make variations

$$\mathbf{u}_i \pm \epsilon \varphi \qquad (\varphi \geq 0, \qquad \epsilon > 0).$$

Therefore each u_i satisfies a PDE in the region $\{|\mathbf{u}| > 0\}$. On the boundary $\partial\{|\mathbf{u}| > 0\}$, only variations upward $\mathbf{u}_i + \epsilon \varphi$.

For $f(\mathbf{u}) = 2|\mathbf{u}|$, one obtains a PDE of the type

$$\Delta \mathbf{u} = \mathbf{g}(x) \chi_{\{|\mathbf{u}| > 0\}}, \qquad \text{where} \ \ \mathbf{g}(x) = \frac{\mathbf{u}}{|\mathbf{u}|}.$$

For the scalar case this corresponds to Obstacle problem $\Delta u = \chi_{\{u > 0\}}$.
For $f(\mathbf{u}) = \chi_{\{|\mathbf{u}| > 0\}}$, one obtains a PDE of the type

$$\Delta \mathbf{u} = \mathbf{w} \mathscr{H}^{n-1} \llcorner (\Omega \cap \partial^*\{|\mathbf{u}| > 0\}),$$

where

$$\mathbf{w}(x) = \lim_{y \in \{|\mathbf{u}| > 0\}, y \to x} \frac{\mathbf{u}(y)}{|\mathbf{u}(y)|}.$$

This corresponds to Bernoulli problem

$$|\nabla \mathbf{u}|^2 = 1 \quad \text{on } \partial^*\{|\mathbf{u}| > 0\}.$$

Here ∂^* denotes the smooth (reduced) part of the boundary.
   A main question that may arise is: Can we reduce the PDE-s to scalar case?

Can we show that for some $\alpha > 0$ $\mathbf{g}, \mathbf{w} \in C^\alpha$?

If so, then we fall back to the scalar case for one of u_i. Hence we can apply known methods for scalar case, and obtain regularity results. Let us see how this can work:

For $h_1, h_2 \geq$ and harmonic in a NTA[22] domain D, (no-cusps) with $h_1 = h_2 = 0$ on $\partial D \cap B_{1/2}(z)$ ($z \in \partial D$) one has

$$\frac{h_1}{h_2} \in C^\alpha(D \cap B_r(z)), \quad \text{some } r > 0.$$

If we could apply this to our system with $h_1 = u_i$, for any $i = 1, \cdots, m - 1$, and $h_2 = u_m$ (assuming u_m is non-degenerate) then for $i = 1, \cdots, m - 1$

$$A_i := \frac{u_i}{u_m} \in C^\alpha \qquad \text{near the free boundary.}$$

Hence,

$$\frac{u_m}{|\mathbf{u}|} = \frac{1}{\sqrt{A_1 + \cdots + A_{m-1} + 1}} \in C^\alpha.$$

But this is exactly the term g_m, or w_m in the problems above. Hence we can apply classical scalar theory. This has successfully been done for the second problem where $f(\mathbf{u}) = \chi_{|\mathbf{u}|>0}$ in [9].

The above system may also be seen as one of the simplest extensions of the classical *obstacle problem* to the vector-valued case: Solutions of the classical obstacle problem are minimisers of the energy

$$\int_D (\frac{1}{2}|\nabla u|^2 + \max(u, 0)) \, dx,$$

where $u : \mathbb{R}^n \supset D \to \mathbb{R}$.

In the scalar case ($m = 1$), one recovers the two phase free boundary problem

$$\Delta u = \chi_{\{u>0\}} - \chi_{\{u<0\}} \quad (= \frac{u}{|u|}),$$

which is a well-studied problem.

For the case $m = 2$ we give an example (due to Nina Uraltseva) that illuminates the possible difficulties of the problem. Consider both real and imaginary parts of the function

$$S(z) = z^2 \log |z| \qquad (z = x + iy)$$

---

[22]NTA domains are domains that satisfy the corkscrew condition, and the Harnack Chain Condition. Examples of NTA domains: Lipschitz domains, domains with Boundary that is graph of a function with gradient in BMO, Chord arc domains (snow-flakes).

which satisfy the **unstable** equation (up to a multiplicative constant) and have singularities at the origin:

$$\Delta u_i = \frac{-u_i}{|\mathbf{u}|}, \quad i = 1, 2, \ldots$$

Hence optimal $C^{1,1}$ regularity is lost for the unstable problem!

To give an idea of how solutions may look like we give a few simple examples of solutions

1. $u_i = \alpha_i P(x)$, with $P(x) \geq 0$, $\Delta P(x) = 1$, and $\sum_{i=1}^{m} \alpha_i^2 = 1$,
2. $u_i = \alpha_i (x_1^+)^2/2 + \beta_i (x_1^-)^2/2$, (2-phase)
   $\sum_{i=1}^{m} \alpha_i^2 = 1$, $\sum_{i=1}^{m} \beta_i^2 = 1$,
3. $u_i = \alpha_i (x_1^+)^2/2$, (1-phase) $\sum_{i=1}^{m} \alpha_i^2 = 1$.

We shall be interested in the class of solutions $\mathbf{u}$ that asymptotically, near a free boundary point, behave like

$$\frac{\max(x \cdot \nu, 0)^2}{2} \epsilon$$

where ν is a unit vector in $\mathbb{R}^n$ and ϵ is a unit vector in $\mathbb{R}^m$.

Also denote by $\mathbb{H}$ the class of all these Half-space solutions.

One can work out many properties of the solution $\mathbf{u}$ to our problem:

- **Uniqueness:** use first variation by $\phi := \mathbf{u} - \mathbf{v}$, both having the same boundary data.
- **Bounds:** $\sup_{B_{3/4}} |\mathbf{u}| + \sup_{B_{3/4}} |\nabla \mathbf{u}| \leq C_1(n, m) \left(\|\mathbf{u}\|_{L^1(B_1; \mathbb{R}^m)} + 1 \right)$.
- **Stability:** $\mathbf{u}_k \to \mathbf{u}$ weakly in $W^{1,2}(D; \mathbb{R}^m)$ then Rellich's theorem together with the fact that $D^2\mathbf{u} = 0$ a.e. in $\{\mathbf{u} = 0\}$, implies that $\mathbf{u}$ is a solution, too.

- **Non-Degeneracy:**

$$\sup_{B_r(x^0)} |\mathbf{u}| \geq \frac{1}{2n}r^2, \quad \forall x^0 \in \overline{\{|\mathbf{u}| > 0\}}.$$

Use the fact that $\Delta|\mathbf{u}| \geq 1$ in the set $\{|\mathbf{u}| > 0\}$.
- **L^1-closeness implies geometric closeness:** If $\|\mathbf{u} - \mathbf{h}\|_{L^1(B_1; \mathbb{R}^m)} \leq \epsilon < 1$, where $\mathbf{h} := \frac{\max(x_n, 0)^2}{2} \epsilon^1$. Then

$$B_{1/2}(0) \cap \operatorname{supp} \mathbf{u} \subset \left\{ x_n > -C\epsilon^{\frac{1}{2n+2}} \right\}$$

with a constant $C = C(n, m)$.

We define a new energy functional

$$\mathcal{M}(\mathbf{v}, x^0, r) := \frac{1}{r^{n+2}} \int_{B_r(x^0)} (|\nabla \mathbf{v}|^2 + 2|\mathbf{v}|) - \frac{2}{r^{n+3}} \int_{\partial B_r(x^0)} |\mathbf{v}|^2 \, d\mathcal{H}^{n-1},$$

which will be used to prove both the growth of $\mathbf{u}$ from the free boundary and also the behavior of the free boundary at good points.

For $\mathbf{v}_r(x) := r^{-2}\mathbf{v}(rx + x^0)$, we have

$$\mathcal{M}(\mathbf{v}, x^0, r) = \mathcal{M}(\mathbf{v}_r, 0, 1) =: \mathcal{M}(\mathbf{v}_r),$$

and $\mathcal{M}$ is monotone in r:

$$\frac{d\mathcal{M}(\mathbf{v}, x^0, r)}{dr} \geq 0.$$

We also define

$$\frac{\alpha_n}{2} := \mathcal{M}\left(\frac{\max(x \cdot \nu, 0)^2}{2} \epsilon\right),$$

and one can show that $\alpha_n = 2\mathcal{M}(\mathbf{h})$ for all $\mathbf{h} \in \mathbb{H}$.

- $\mathcal{M}(\mathbf{u}) \geq \frac{\alpha_n}{2}$ for all 2-homogeneous global solutions, and with equality if and only if $\mathbf{u} \in \mathbb{H}$.
- In the $L^1(B_1(0); \mathbb{R}^m)$- topology, $\mathbb{H}$ is isolated within the class of homogeneous solutions of degree 2.

We define

$$\Gamma(\mathbf{u}) := D \cap \partial\{x \in D : |\mathbf{u}(x)| > 0\},$$

$$x \in \Gamma_0(\mathbf{u}) := \Gamma(\mathbf{u}) \cap \{x : \nabla \mathbf{u}(x) = 0\}$$

A point x is a regular free boundary point for $\mathbf{u}$ if:

$$x \in \Gamma_0(\mathbf{u}) \quad \text{and} \quad \lim_{r \to 0} \mathcal{M}(\mathbf{u}, x, r) = \frac{\alpha_n}{2}.$$

We denote by $\mathcal{R}_u$ the set of all regular free boundary points of u in B_1.

From the upper semicontinuity of $\mathcal{M}(\mathbf{u}, x, r)$ in $\mathbf{u}$, and the isolated property of $\mathbb{H}$ we can conclude that the set of regular free boundary points $\mathcal{R}_u$ is open relative to $\Gamma_0(\mathbf{u})$.

Any solution $\mathbf{u}$ to our system in $B_1(0)$ satisfies

$$|\mathbf{u}(x)| \leq C\mathbf{dist}^2(x, \Gamma_0(u))$$

and

$$|\nabla \mathbf{u}(x)| \leq C \mathbf{dist}(x, \Gamma_0(\mathbf{u})) \text{ for every } x \in B_{1/2}(0),$$

where the constant C depends only on n and

$$E(\mathbf{u}, 0, 1) = \int_{B_1(0)} (|\nabla \mathbf{u}|^2 + 2|\mathbf{u}|).$$

Using the monotonicity formula one can show (elementary)

$$\frac{2}{r^{n+2}} \int_{B_r} |\mathbf{u}| \leq E(\mathbf{u}, 0, 1) + \frac{2}{r^{n+3}} \int_{\partial B_r} |\mathbf{u} - \mathbf{p}|^2 d\mathcal{H}^{n-1}$$

$$\leq C_0 + C_1(\mathbf{p})|D^2 \mathbf{u}|_{BMO(B_{1/2})} \leq C_2$$

for each $\mathbf{p} = (p_1, \ldots, p_m)$ such that each component p_j is a homogeneous harmonic polynomial of second order.

The regularity of the free boundary follows through an epiperimetric inequality, which goes as follows: There exists $\kappa \in (0, 1)$ and $\delta > 0$ such that if $\mathbf{c}$ is a homogeneous function of degree 2 satisfying $\|\mathbf{c} - \mathbf{h}\|_{W^{1,2}(B_1;\mathbb{R}^m)} + \|\mathbf{c} - \mathbf{h}\|_{L^\infty(B_1;\mathbb{R}^m)} \leq \delta$ for some $\mathbf{h} \in \mathbb{H}$, then there is a $\mathbf{v} \in W^{1,2}(B_1; \mathbb{R}^m)$ such that $\mathbf{v} = \mathbf{c}$ on ∂B_1 and

$$\mathscr{M}(\mathbf{v}) \leq (1 - \kappa) \mathscr{M}(\mathbf{c}) + \kappa \frac{\alpha_n}{2}.$$

In lay terms this says that

$$|\mathscr{M}(\mathbf{u}_{r_1}) - \mathscr{M}(\mathbf{u}_{r_2})| \leq c|r_2 - r_1|^\alpha, \qquad \alpha = \alpha_\kappa,$$

and gives uniqueness of the blow-ups.

One shows there exist $\beta' > 0$, $r_0 > 0$ and $C < \infty$:

$$\int_{\partial B_1(0)} \left| \frac{\mathbf{u}(x^0 + rx)}{r^2} - \frac{1}{2}\epsilon(x^0) \max(x \cdot v(x^0), 0)^2 \right| d\mathcal{H}^{n-1} \leq C r^{\beta'}$$

for every $x^0 \in \mathscr{R}_u$ and every $r \leq r_0$.

Here $v(x^0)$ depends on the blow-up of $\mathbf{u}$ at x^0. This implies that

$$\boxed{\mathscr{R}_u \text{ is locally in } D \text{ a } C^{1,\beta}\text{-surface.}}$$

Acknowledgement H. Shahgholian has been supported in part by Swedish Research Council.

References

1. J. Andersson, H. Mikayelyan, The zero level set for a certain weak solution, with applications to the Bellman equations. Trans. Am. Math. Soc. **365**(5), 2297–2316 (2013)
2. J. Andersson, H. Shahgholian, G. Weiss, A variational linearization technique in free boundary problems applied to a two-phase bernoulli problem. Manuscript
3. R. Barkhudaryan, D.A. Gomes, M. Salehi, H. Shahgholian, System of variational inequalities with obstacles. Manuscript
4. I. Blank, Alexander Sharp results for the regularity and stability of the free boundary in the obstacle problem. Thesis (Ph.D.), New York University, 2000, 56 pp.
5. L.A. Caffarelli, A remark on the Hausdorff measure of a free boundary, and the convergence of coincidence sets. (Italian summary). Boll. Un. Mat. Ital. A (5) **18**(1), 109–113 (1981)
6. L.A. Caffarelli, The obstacle problem revisited. J. Fourier Anal. Appl. **4**(4–5), 383–402 (1998)
7. L.A. Caffarelli, U. Stefanelli, A counterexample to $C^{2,1}$ regularity for parabolic fully nonlinear equations. Commun. Partial Differ. Equ. **33**(7), pp. 1216–1234 (2008)
8. L.A. Caffarelli, L. Karp, H. Shahgholian, Regularity of a free boundary with application to the Pompeiu problem. Ann. Math. (2) **151**(1), 269–292 (2000)
9. L. Caffarelli, H. Shahgholian, K. Yeressyan, A minimization problem with free boundary related to a cooperative system. Duke Univ. J. (to appear)
10. P. Diaconis, W. Fulton, A growth model, a game, an algebra, Lagrange inversion, and characteristic classes. (English summary) Commutative algebra and algebraic geometry, II (Italian) (Turin, 1990). Rend. Sem. Mat. Univ. Politec. Torino **49**(1), 95–119 (1991/1993)
11. A. Figalli, H. Shahgholian, A general class of free boundary problems for fully nonlinear elliptic equations. Arch. Ration. Mech. Anal. **213**(1), 269–286 (2014)
12. A. Figalli, H. Shahgholian, A general class of free boundary problems for fully nonlinear parabolic equations. Ann. Mat. Pura Appl. (4) **194**(4), 1123–1134 (2015)
13. M. Fotouhi, H. Shahgholian, A semilinear PDE with free boundary. Nonlinear Anal. **151**, 145163 (2017)
14. A. Friedman, *Variational Principles and Free-Boundary Problems*. Pure and Applied Mathematics (A Wiley-Interscience Publication/Wiley, New York, 1982), ix+710 pp. ISBN: 0-471-86849-3
15. B. Gustafsson, H.S. Shapiro, What is a quadrature domain? in *Quadrature Domains and Their Applications*. Operator Theory, Advances and Applications, vol. 156, 125 (BirkhŁuser, Basel, 2005)
16. B. Gustafsson, A. Vasil'ev, *Conformal and Potential Analysis in Hele-Shaw Cells*. Advances in Mathematical Fluid Mechanics (Birkhäuser Verlag, Basel, 2006), x+231 pp. ISBN: 978-3-7643-7703-8; 3-7643-7703-8
17. H.S. Hele-Shaw, On the motion of a viscous fluid between two parallel plates. Trans. R. Inst. Nav. Archit. London **40**, 21 (1898)
18. S. Kim(KR-SNU), K.-A. Lee(KR-SNU), H. Shahgholian(S-RIT), An elliptic free boundary arising from the jump of conductivity. (English summary) Nonlinear Anal. **161**, 129 (2017)
19. S. Kim(KR-SNU), K.-A. Lee(KR-SNU), H. Shahgholian(S-RIT), Nodal sets for "Broken" quasilinear PDEs. Manuscript
20. K. Lee, J. Park, H. Shahgholian, The regularity theory for thin double obstacle problem. Manuscript
21. S.A. Molchanov, E. Ostrovskii, Symmetric stable processes as traces of degenerate diffusion processes. Theory Probab. Appl. **14**, 128–131 (1969)
22. A. Petrosyan, H. Shahgholian, N. Uraltseva, *Regularity of Free Boundaries in Obstacle-Type Problems*. Graduate Studies in Mathematics, vol. 136 (American Mathematical Society, Providence, 2012).
23. G.S. Weiss, A homogeneity improvement approach to the obstacle problem. Invent. Math. **138**(1), 2350 (1999)

LECTURE NOTES IN MATHEMATICS  Springer

Editors in Chief: J.-M. Morel, B. Teissier;

Editorial Policy

1. Lecture Notes aim to report new developments in all areas of mathematics and their applications – quickly, informally and at a high level. Mathematical texts analysing new developments in modelling and numerical simulation are welcome.

   Manuscripts should be reasonably self-contained and rounded off. Thus they may, and often will, present not only results of the author but also related work by other people. They may be based on specialised lecture courses. Furthermore, the manuscripts should provide sufficient motivation, examples and applications. This clearly distinguishes Lecture Notes from journal articles or technical reports which normally are very concise. Articles intended for a journal but too long to be accepted by most journals, usually do not have this "lecture notes" character. For similar reasons it is unusual for doctoral theses to be accepted for the Lecture Notes series, though habilitation theses may be appropriate.

2. Besides monographs, multi-author manuscripts resulting from SUMMER SCHOOLS or similar INTENSIVE COURSES are welcome, provided their objective was held to present an active mathematical topic to an audience at the beginning or intermediate graduate level (a list of participants should be provided).

   The resulting manuscript should not be just a collection of course notes, but should require advance planning and coordination among the main lecturers. The subject matter should dictate the structure of the book. This structure should be motivated and explained in a scientific introduction, and the notation, references, index and formulation of results should be, if possible, unified by the editors. Each contribution should have an abstract and an introduction referring to the other contributions. In other words, more preparatory work must go into a multi-authored volume than simply assembling a disparate collection of papers, communicated at the event.

3. Manuscripts should be submitted either online at www.editorialmanager.com/lnm to Springer's mathematics editorial in Heidelberg, or electronically to one of the series editors. Authors should be aware that incomplete or insufficiently close-to-final manuscripts almost always result in longer refereeing times and nevertheless unclear referees' recommendations, making further refereeing of a final draft necessary. The strict minimum amount of material that will be considered should include a detailed outline describing the planned contents of each chapter, a bibliography and several sample chapters. Parallel submission of a manuscript to another publisher while under consideration for LNM is not acceptable and can lead to rejection.

4. In general, **monographs** will be sent out to at least 2 external referees for evaluation.

   A final decision to publish can be made only on the basis of the complete manuscript, however a refereeing process leading to a preliminary decision can be based on a pre-final or incomplete manuscript.

   Volume Editors of **multi-author works** are expected to arrange for the refereeing, to the usual scientific standards, of the individual contributions. If the resulting reports can be forwarded to the LNM Editorial Board, this is very helpful. If no reports are forwarded or if other questions remain unclear in respect of homogeneity etc, the series editors may wish to consult external referees for an overall evaluation of the volume.

5. Manuscripts should in general be submitted in English. Final manuscripts should contain at least 100 pages of mathematical text and should always include

   - a table of contents;
   - an informative introduction, with adequate motivation and perhaps some historical remarks: it should be accessible to a reader not intimately familiar with the topic treated;
   - a subject index: as a rule this is genuinely helpful for the reader.
   - For evaluation purposes, manuscripts should be submitted as pdf files.

6. Careful preparation of the manuscripts will help keep production time short besides ensuring satisfactory appearance of the finished book in print and online. After acceptance of the manuscript authors will be asked to prepare the final LaTeX source files (see LaTeX templates online: https://www.springer.com/gb/authors-editors/book-authors-editors/manuscriptpreparation/5636) plus the corresponding pdf- or zipped ps-file. The LaTeX source files are essential for producing the full-text online version of the book, see http://link.springer.com/bookseries/304 for the existing online volumes of LNM). The technical production of a Lecture Notes volume takes approximately 12 weeks. Additional instructions, if necessary, are available on request from lnm@springer.com.

7. Authors receive a total of 30 free copies of their volume and free access to their book on SpringerLink, but no royalties. They are entitled to a discount of 33.3 % on the price of Springer books purchased for their personal use, if ordering directly from Springer.

8. Commitment to publish is made by a *Publishing Agreement*; contributing authors of multiauthor books are requested to sign a *Consent to Publish form*. Springer-Verlag registers the copyright for each volume. Authors are free to reuse material contained in their LNM volumes in later publications: a brief written (or e-mail) request for formal permission is sufficient.

Addresses:
Professor Jean-Michel Morel, CMLA, École Normale Supérieure de Cachan, France
E-mail: moreljeanmichel@gmail.com

Professor Bernard Teissier, Equipe Géométrie et Dynamique,
Institut de Mathématiques de Jussieu – Paris Rive Gauche, Paris, France
E-mail: bernard.teissier@imj-prg.fr

Springer: Ute McCrory, Mathematics, Heidelberg, Germany,
E-mail: lnm@springer.com

Printed in the United States
By Bookmasters